Science for development

Jacques Spaey
with the collaboration of
Jacques Defay,
Jean Ladrière,
Alain Stenmans
and Jacques Wautrequin

Science
for development

An essay on the origin
and organization
of national science policies

Unesco Paris 1971

The translation of this essay into English
was carried out by the Science Policy Division
of the Unesco Secretariat

Published by the United Nations
Educational, Scientific and Cultural Organization,
Place de Fontenoy, 75 Paris-7e

Printed by Snoeck-Ducaju et Fils, Gand

Preface

A major objective of the Unesco science policy programme is to promote more active intellectual co-operation among decision-makers and specialists concerned with the symbiosis science-government, for what hangs upon this symbiosis, in the world of today, is not merely social and economic progress, but indeed the very security of nations. This Unesco programme also seeks to help Member States, at their request, to set in motion and to promote throughout the national economy a process of endogenous development, innovative in character, based on the application of modern science and technology.

In this essay—written by a team of specialists under the direction of Jacques Spaey, Secretary General of the Prime Minister's Department for the programming of science policy in Belgium—the authors analyse the currents of thought and action which characterize national science policies among the European States, and outline many of the pressing problems currently facing those responsible for governmental development policies based on science. The essay was published in its original French version in 1969.

The opinions expressed by the authors commit no one but themselves. Unesco feels that the world-wide dimension which the authors have given to their argument, and the positive form in which they advance it, will earn them attentive readers among all those working for the cause of co-operation between the nations.

Dr. Jacques Spaey, leader of the team of authors responsible for the present publication, died shortly before the appearance of this English edition.

Throughout the international community, Dr. Spaey was held in high esteem for the skill and imagination with which he sought to harness science and technology to man's well-being, and for his constant efforts to further the principles of co-operation between the nations.

The Secretariat of Unesco pays tribute to his memory.

Contents

Introduction

This essay is, in the full sense of the word, the work of a team. It was conceived and carried out by a group of which the authors are the spokesmen but not the only artisans. Each member of the team has, on his own account, devised, nurtured and sustained the flow of ideas which has led the team to a common approach.

Together they have overcome the obstacles which their activity encountered, for every effort towards progress necessarily entails change or new orientations of existing institutions or accepted ideas. Participation in the shaping of the authors' views has extended to different circles interested in the promotion of science and technology. Their advice and their criticism have given scope for a confrontation of current ideas with the realities of today and tomorrow.

Furthermore, the studies which lie at the basis of this essay have been debated on the international plane. The exchange of views carried out in international organizations, more particularly in Unesco and OECD, on the science policy of governments, has made it possible to apprehend many ways in which science and technology contribute to the development of human society. Such confrontations of experience and ideas, originating in different countries, have made it possible to generalize some of the conclusions and to identify methods applicable in a large number of countries.

A firm hope of progress, and the will to overcome the present constraints on development, have inspired our activity as much as our thinking. Thus our initial assumption was that science, while drawing its inspiration from men's curiosity, has become at the same time an essential factor in development and progress.

In countries where the promotion of science remains but marginal to economic and social development policies, the rhythm and quality of progress suffers, just as its dissemination does, from serious delays. Conversely, in the highly developed nations, it is the systematic application of scientific knowledge which has spurred on technological innovation, with its far-reaching consequences for economic and social progress. The fact that the systematic use of science has often been concentrated on military or political objectives

11

should not lead us to underestimate its potentialities for application to peaceful ends.

A number of observations highlighted in this essay suggest that the incursion of science and technology into all human activities imparts new dimensions and higher rates of occurrence to the problems raised by the development of societies. It will soon be possible to test this hypothesis, since many countries are embarking on a process of systematic organization of their scientific and technological activities. Experience thus gained will allow for adjustments to be made in the interpretations and approaches which are proposed in this essay.

Our purpose has been to stress the above-mentioned observations and, if one prefers to look at it in another way, to support the underlying working hypothesis. In the final reckoning the essay deals with a theory of *development based on science,* whose conditions and limits we shall delineate as they appear today.

Thus we hope to make a contribution to some essential aspects of 'science policy' and to provide those who assume the responsibility for it in their respective countries with arguments and methods for undertaking vigorous action in this field, where the progress of each nation depends on the comparison of experiences and the co-operation of all.

As we have said, this essay is the work of a team; it is therefore difficult to give a name to each individual contribution. However, it must be emphasized that Professor J. Ladrière has brought us an original contribution on the meaning of scientific and technological change in the thought and organization of contemporary society. Mr. J. Defay has devoted himself more particularly to defining the place of scientific activities in the economic and social context and, at the same time, to the part played by science in development. Mr. A. Stenmans has applied himself to the definition of the functions of scientific policy on the political and administrative plane. Mr. J. Wautrequin has described the birth and evolution of the organized planning of science policy in different countries, the machinery for promoting research, and the problems of scientific co-operation in the international field.

Finally our thanks go to those who, without having shared directly in the drafting of the different chapters of this work, have contributed to it in various ways. Mr. J. De Meulder and Mr. J. Sommeryns have made a summary of the survey techniques and budgetary analysis which are at the basis of a science policy. Miss G. Dehoux, who is in control of administration, has co-ordinated the groundwork which made this publication possible.

The plan of this essay has been conceived in such a way as to give to the reader an alternation between concrete and theoretical information, which allows him to become progressively familiar with a somewhat abstruse subject.

First part. The first chapter summarizes the objectives that are usually associated with a national policy of economic and social development. It enlarges a little on economic growth, which clearly occupies a central place in the objectives of such a policy. However, it gives these concerns their proper place among the more general purposes which form the aims of civilization. The second chapter deals with scientific and technological change, and the demands of rationality and logic which are its distinguishing features. It describes the scientific mutation of societies as a voluntary action of a new kind, in which man sets the final goals of his activities.

Second part. The third chapter describes the actual state of affairs which constitutes the given data and therefore defines the point of departure of government action. National situations are infinitely variable. An attempt at classification has therefore been made with the object of identifying them according to the state of development already reached by the nation at the moment when the programme of development based on science is conceived, decided upon and put into operation.

We are fully conscious of the arbitrary nature of any kind of classification, and our only aim in this chapter has been to propose a method of analysis of a country's situation. The results of such an analysis allow the identification and subsequent choice of certain objectives of national development.

In the fourth chapter we come into contact with certain realities of scale. By enabling us to appreciate the extent of the size and scope of the scientific revolution in the world of today, they reveal the imbalances which must be faced if the world's evolution is not to take a disruptive course. At the same time they show the range of the key parameters and variables of which the specialist in scientific programming must take account and avail himself.

The fifth chapter summarizes the methods and organizations of the past for the promotion of research, and shows how today's science policies have gradually taken shape and substance. This historical survey ends the examination of the essential facts which must be taken account of in the formulation of a development policy based on science.

Third part. The sixth chapter gives science policy its place as part of governmental policy, while the seventh chapter attempts a definition of its functions and machinery.

Lastly, the eighth chapter deals with scientific and technical co-operation between the nations.

Jacques Spaey

Part one

The role and dynamics of science in contemporary society

1 Science, a means of development

INTRODUCTION

SCIENCE AND PROGRESS

Knowledge for 'action'. This is without doubt the main trait of contemporary civilization.

The pursuit of knowledge for its own sake remains certainly one of the noblest and most creative motivations of scientists throughout history. But today it tends to be jostled out by the urge for application. Men of today believe in 'efficiency'. They consider it as an imperative duty when the necessary conditions of success have come in sight. They set out to prepare such conditions carefully, while turning their backs on fatalism and inaction. Ours is a problem-solving civilization.

However, it would be a mistake to think that the ethic underlying this action-oriented behaviour which has spread over the world is necessarily and solely utilitarian. Great adventures like the conquest of the Moon and the planets are 'economically unproductive'.[1] If, nevertheless, resources have been found to enter into the space age, it is because astronautics possesses a value of its own: it provides mankind with an opportunity and a challenge. The national prestige which redounds from space travel reflects the honour that all men feel in their hearts for exceptional accomplishments which extend human experience.

Of course, the importance of the part played by power and profit-motivated policies should not be underestimated. But it should be pointed out that 'action' has become an end in itself in our scale of values, and is therefore a motivation as powerful as the search for truth and the advancement of

1. Nevertheless, the National Aeronautics and Space Administration (NASA) rightly attaches much importance to the favourable economic consequences of space research (the 'fall-out' and the 'spill-over' effects). This aspect of astronautics is likely to make other countries or groups of countries engage in space exploration. It remains true that economic returns of space navigation are not among the prime motives for the American and Soviet decisions; nor are these purely military.

17

knowledge. Professor J. Ladrière in the second chapter of this essay even suggests that 'action' allows man to discover, to fulfil and to justify his ultimate destiny.

But whether his motives are utilitarian or not, man needs more knowledge in order to act effectively. Thus, the will to action feeds and nourishes the thirst for knowledge.

It has not always been so. For a long time, knowledge remained a speculative activity, while action was essentially empirical. Science gave no help to the craftsman, the architect, the sailor, the strategist or the doctor in overcoming their difficulties, any more than the empirical observation of the technician helped the philosopher to understand the world. This separation goes far to explain the slow pace of technological progress in the past centuries, while the recent alliance of theoretical knowledge and empirical observation accounts for its sudden acceleration.

CONTROL OF THE NATURAL ENVIRONMENT

Science nowadays also appears as a means to be used in the pursuit of definite goals. Certainly, the strategists and the industrialists learnt long ago how to make use of scientific discoveries and technical inventions. But, to base plans of action on truths as yet untested and on technical feats not ascertained by experience is indeed the daring distinction of our epoch. It is in the field of industrial competition, and afterwards in that of military rivalry, that 'science-based policies' first appeared. These are now widespread, for none of the great military powers or important industrial groups would today dare run the risk of being caught unprepared by the technological breakthroughs which are being achieved in the research and experimental development laboratories of their leading competitors.

Nations or industrial enterprises which conduct large and well-organized research and experimental development operations can nowadays bank on decisive results. Furthermore, it is known that the probability of success increases with the size of the research effort, thanks to the statistical spreading of the risks. It grows also with the quality of the strategy, and that of the economic and technological forecasting on which the planning of research programmes is founded.

If science is thus harnessed for competitive rivalries between firms and between nations, it can also be utilized to surmount the natural obstacles which block the path of social groups to prosperity and development. And why should it not find its place also in the plans of action which the world community makes for its own benefit? The reclamation of deserts, and the solution of such ills as the degradation of soils, plague, illness, the pollution of air and water, are today the subject of major co-operative plans of action. It is indeed deplorable that fewer scientific resources have been made available for such social goals than those mobilized for the needs of industrial competition or defensive strategy. At the same time, it should be recognized that the latter have provided the driving force for spectacular progress in science and technology, which has become available for peaceful ends.

THE SCIENTIFIC STUDY OF MAN AND OF HUMAN SOCIETY

The searching mind of man, and his scientific approach to problems, have made the natural environment more intelligible and at the same stroke have increased tenfold the power of mankind over the material universe. This same effort should now be turned upon the biosphere, of which man is a part. Man himself would thus become, more than ever before, an object of scientific study. Knowledge in this field, as in all others, increases the power of action.

All action requires an acting subject, an object to be acted upon, and the means to perform the action. Once man himself is chosen as the object of the action under consideration, an ethical problem arises, for the object is no longer one selected from the material universe. The respect due to man, when he becomes the object of the action, is the first and obvious limitation.

But the means to perform the action can be equally well a human person or a group of persons. In the eyes of the acting subject and with reference to his objective, this person (or this group) is a tool which he must put into operation. In the first stage of the action, the human means undergoes a kind of ethical mutation: it is brought down to the level of an instrument with reference to the whole action under consideration. Thus in the use of another human being, there appears at the second stage the overpowering constraint of respect for the human being.

The ethical problem of the human use of human beings is certainly not new. It has occurred since the earliest stages of man's history. It has emerged under a different aspect every time man's power has been increased by technical inventions, and also when the operational specialization of the individual's functions in society has removed him further from the appreciation of the ends of his actions, thus making him at each stage more 'object' and less 'subject'. The powerful means of action derived from modern science confront our generation with this acute ethical problem on a planetary or even a cosmic scale.

It is evidently necessary that limits should be fixed to the actions by which the individual or society can suffer being made use of or manipulated. This is the negative aspect of the ethical problem. It is equally necessary to choose the ends for which the utilization of the new power of action can be put into operation. This is its positive aspect. Society cannot escape the necessity of defining these ends from the moment when its future ceases any longer to be shaped merely by the play of unintelligible or uncontrollable natural forces. One of the most tragic constraints of the human condition—and also one of the most noble—is indeed this responsibility which grows proportionately as the thirst for knowledge and the will to power are satisfied. Man simply must shoulder fully his responsibility: indeed, no other choice is worthy of him.

The *civilization blueprint* which it is the duty of contemporary society to sketch out will then be expressed by the *ultimate goals chosen* by men for their collective operations on the natural environment and on the biosphere (including society and man himself), and by the *limits set* to these actions.

THE CHOICE OF THE COLLECTIVE 'CIVILIZATION BLUEPRINT'

The objectives of nations and governments have been for a long time as simple as they are obvious: to survive, and if possible to conquer, in the competitive struggle with other nations; to face up to poverty of resources and to natural disasters; to keep oneself on the path of progress.

In the course of the last centuries and especially of the last decades, more ambitious and complex ultimate goals have appeared which include in particular the building of a different society, the deliberate speeding up of development, or again the quest for foreign relations founded on the integration and co-operation of nations rather than on their rivalry.

It is true that few nations or governments have consciously drawn up their 'civilization blueprint'. But all find themselves today put under the pressure of unsatisfied aspirations, and have been led gradually to formulate the ultimate goals which they pursue and the means which they choose to attain them.

In fact the masses have become conscious of the universal possibilities which the scientific approach offers. It proceeds from knowledge ascertained by experiment towards the creation of the tool, the model or the methodology, and from that to the planned setting into operation of the means thus created. Even when it is a question of natural disasters like floods or drought, the governments can no longer appeal solely to fate, for the peoples have forgotten how to curse heaven for the ills which overwhelm them or the hopes in which they have been frustrated. From now on, it is of their governments that they demand reckoning. The power which has not provided itself with the means of action, or which has not made use of those means, finds itself liable to the reproach of incapacity, of bad policy or lack of energy. Any one of these condemns it in the eyes of young people brought up in an action-oriented ethic and the cult of technical efficiency.

This is why all governments are led in the end, either through their own initiative or through the impatience of their people, to formulate development goals for their country, to define them in concrete terms, and to make a plan of action to achieve them. Such a plan calls not only for scientific knowledge of the object of the action (natural environment, man and society) but also for the creation of new techniques and new systems of action. Research here intrudes under two essential headings: the knowledge of the object, and the creation of the means of action.

Among the collective 'civilization blueprints' which the nations are led to choose for themselves are found those which are nakedly materialistic, such as the increase of the total of goods and services available per head of the population.

There are others of a more idealistic nature. These are the ones which concern the development of man himself, the functional advances of social organization, and the ethical change which the individual and society have to accomplish in order to prevent the growth of their power from turning to their own detriment. These more idealistic goals concern the quality of life, both individual and social.

The objective of economic growth is nevertheless everywhere present. No contemporary society views its objectives of individual and social progress in the perspective of poverty and a low level of material consumption. All intend, at least in principle, to make use of the added resources which come from the increase in production, in order to make qualitative progress in the life of men, even though the ultimate goal may sometimes be lost sight of because of the attraction of the immediate objectives.

So, as the logic of the argument will oblige the authors of this essay to deal in the first instance with the links between science and economic growth, let it be understood that—at least as far as they are concerned—economic growth cannot be regarded as an ultimate goal in itself.

SCIENCE IN SOCIETY

The connexion which today is accepted between interpretive reasoning and the will to action has existed for hardly more than three or four centuries. It was in the age of the Renaissance that scholars began to accept the submission of their theoretical knowledge to the proof of experiment, and thus to yield to the compulsion of facts. It was in this age also that the arts and crafts began to recognize the superiority of the scientific approach (characterized by the theoretical explanation of phenomena which man wants to make use of, or to protect himself against) over the empirical approach which only makes progress by the endless repetition of trial and error. The union of science and technology is therefore an act of convergence and integration of two attitudes; the scholar has become an experimenter, and the craftsman has become a man of science in seeking to understand phenomena in order to control them. Since this integration has come about—and it did not become complete until the middle of the nineteenth century—the distinction between fundamental research, applied research, and experimental development has lost much of its meaning, so far as the machinery of programming is concerned. Every day it loses more, because all large-scale programmes combine the three stages of the process.

Nevertheless the distinction remains in the analysis, *a posteriori,* of the completed endeavour. Henceforth the distinction rests more on the immediate motivation, than on the nature of, or the method used in, the research. The scientific worker who devotes his energies to *fundamental research* does not know in advance what others, or he himself, will gain—practically speaking—from the new knowledge which is the object of his efforts. This is not his concern. It makes it possible for him to conduct his investigations in any direction where his curiosity takes him.

Applied research, on the other hand, aims at gaining insight into the conditions or the causes of success or failure of such and such ways or means of action. The knowledge of the conditions then makes it possible to criticize and eventually improve existing practices. The pressure of needs certainly provides a powerful motivation for the investigator. But in practice it limits the possibility of exploring all the by-paths which one meets on the way.

Experimental development is a process in which the research worker aims at innovation more than at knowledge and understanding. It gives birth to new materials, new processes or techniques, and new instruments and equipment, or it improves those which exist. Thanks to so-called 'exploratory' applied research, whose object is to understand scientifically the phenomena which one intends to use, the team which embarks on an experimental development is assured that 'it can be done'. It knows also the limits which it must impose on its ambition, and the safety conditions which it must enforce. The uncertainties which exist will be removed by the old empirical method of trial and error. But these will be, whenever possible, the object of a planned climax: rough draft, prototype or laboratory model, working model, full-scale completion.

Fundamental research, whether it is theoretical or experimental, finds itself more and more often at the start of the sequence which has just been described. This has been so with electricity, electromagnetic waves, nuclear energy and with many other phenomena of great importance. At a first stage an unusual phenomenon is the object of experimental observations or has been predicted on the basis of theoretical considerations, without any practical application in mind. The scholarly curiosity and the free approach of the fundamental research worker have therefore presided at the inception of these researches.

The part that fundamental research is called upon to play will certainly get larger as science and technological knowledge penetrate into the kind of phenomena which are outside the range of sensual perception or of simple apparatus. The unfettered attitude of the mind, which characterizes such research, opens up new paths—paths which research directed towards practical application would have overlooked or which, if they had been noticed at all, would have been left on one side as having no apparent connexion with the immediate objectives of the research in hand.

As to those who devote themselves to applied research and to experimental development, they bring to non-oriented research a rich harvest of observations made while developing new apparatus and technologies. Workers active in the field also confront the fundamental sciences with demands for the explanation of anomalies that have been met, of failures, of parasitic phenomena, etc. Their thirst for fundamental knowledge is, in fact, insatiable. They are the prime movers of much of the progress of fundamental science. Thus Kepler borrowed the astronomical observation on the position of the planets from the data gathered by Tycho Brahe for the purposes of the navigators and under the pressure of their needs. The intellectual descent from Kepler to Newton and from Newton to the first successes of the scientific method in explaining the world, illustrates the part played by applied research, and by the thirst for knowledge as a basis for action, in the development of scientific thought in Europe.

The integrated whole of these activities has received the name of 'research and development' (R & D). The expression is semantically unimpressive and not very elegant. Its only advantage is that it is widely accepted. In this essay, the word 'research', when it is used alone, means the entirety of the activities covered by the expression 'R & D'. The word 'development', when used alone, means the economic and social progress of a community.

Besides research and development, two other forms of scientific activity are assuming a growing importance in society. They are the scientific public services, and the scientific approach to decision-making.

The *scientific public services* gather, preserve, transform, make accessible and disseminate observations and scientific data on celestial bodies, on the atmosphere, on the soils and the earth crust, on the sea, on living species and their ecology, on man and human society in all its manifestations, etc. These services also deal with the contents of scientific and technological publications and the ways of obtaining them. Besides this, they submit samples or specimens of every substance, inert or living, natural or man-made, to surveys, standardized tests and examinations, which make it possible to assess or identify them according to the criteria of modern science. In fact, they carry out oriented fundamental research, chiefly of a descriptive character, known formerly by the name of 'Natural Sciences'.

The men of our times, whether they are in research or in a practical profession, have a daily need of such information. Sometimes they collect it for themselves. More and more often they have recourse to the scientific public services; and the quality of their work depends on the quality of those services.

The *scientific approach to decision-making* is one of the most recent forms taken by the activity of scientists. Its object is not knowledge for its own sake, or creativity, or public service. It aims to collect and to process information related to the decisions which must be taken in the management of human communities. Its ambition is to foster a more rational, and consequently a more efficient attitude, than that resulting merely from the casual and empirical selection of scientific and technological knowledge, or from personal information, intuition, reasoning and judgement. It is founded on the conviction that—by means of the appropriate instruments—the capacity of man's intellectual faculties can be stepped up, in much the same way that his physical faculties and sensual perceptions have already been stepped up.

THE FACTORS OF ECONOMIC GROWTH

Economic growth is the primary objective of most governments.

Indeed it has become so obvious that technological progress is the key to prosperity, and that scientific research lies at its root, that the temptation arises to call on science for solutions to economic problems, when other ways and means would be more appropriate. The enthusiasm for a new tool easily induces such exaggerations.

It is proper therefore to analyse this idea of economic growth in order to clarify the conditions which govern the efficient use of scientific research as part of a general policy of economic development.

EXTENSIVE GROWTH

An economy can expand by simple multiplication of the production units, without the individual enterprises increasing in size or modifying their operative technologies, and without any economic reorganization of production taking place. This happened frequently in the past when stretches of virgin land were put under cultivation, or when a traditional urban class of artisan could find, in the expansion of foreign markets, the possibility of creating new employment.

Simple growth then takes place by adding new farms or workshops of the same kind to the existing ones, without changing their organization or equipment. This process is therefore called 'extensive growth'.

It assumes an increase in the working population, which can result from an excess of births over deaths, from the existence of a reserve of previously unemployed manpower, or from immigration. The limit of this kind of growth is obviously reached when the available labour force is fully employed.

GROWTH BY MEANS OF CAPITAL ACCUMULATION
WITHOUT TECHNOLOGICAL CHANGE

When producers are provided with more machinery, or more powerful instruments of production—without, however, this equipment being different, e.g. based on any new technology—the process is called 'intensive growth'. One can, for example, work with two horses instead of one, which allows a man to cultivate more acres, or to work the same acreage more intensively.

The limit of this kind of growth is reached when any new increase in production ceases to exceed the additional inputs needed to effect it (e.g. the consumption of fixed capital, that is, interest on capital and depreciation in capital equipment, and the consumption of goods and essential services, such as raw materials, energy, etc.).

The saturation of the economy in capital equipment is reached at this moment. Naturally this saturation only exists in relation to the current operative technologies, which we are assuming to remain fixed.

GROWTH BY MEANS OF ADVANCES IN THE ORGANIZATION OF PRODUCTION
AT CONSTANT TECHNOLOGY AND CAPITAL OUTLAY

When the population and land available are completely committed to production, and when the enterprises have reached the saturation point in capital equipment, it is often still possible to increase productivity by eliminating less productive labour and cutting down delays by, for example, modifying the number and size of the enterprises, and also by changing the way in which labour is organized. The limit of this kind of growth is reached when the use of the available labour force has been fully optimized.

The enlargement of the production units brings reductions in cost price, called 'economies of scale'. These are often obtained from market expansion resulting from political action, or from a lowering of the costs of transport, or again from an increase in the average income of the population. In this last case, the growth is self-sustained. It is in the sector of durable consumer goods, where economies of scale are very important, that the influence of this factor is, for the moment, most evident.

The transfer of manpower from sectors of low productivity to the sectors of high productivity is another kind of change in the structure of the economy which makes important growth possible without technical change. In these sectors, Edward F. Denison[1] has been able to measure the growth potential which many countries still harbour in the excessive allocation of manpower to traditional sectors of the economy. Such over-employment is common in the agriculture sector as well as in some branches of industry, and in the services where family enterprises or handicrafts predominate (self-employment). The shifting of the excess labour force from these sectors to others increases the national product.

It will be noticed that the kinds of growth so far described do not make calls on science as a means or instrument of economic progress. It is not the same in what follows.

GROWTH RESULTING FROM TECHNOLOGICAL INNOVATION
AND ITS PROPAGATION IN THE NATIONAL ECONOMY

The advancement of knowledge makes it possible to devise other means of obtaining the same products or of inventing other products or other machines to satisfy the same needs. A new technology is truly progressive when it absorbs less resources (human labour, capital equipment, intermediate products, and ancillary services) for every unit produced. Technological progress is the essential source of sustained growth in those economies which enjoy at the same time full employment, maximal utilization of capital outlay, and the best possible structure and organization of their production. It is likely that no economy really finds itself in this enviable situation, but some can approach it to the point of becoming chiefly indebted to technological change for their further growth.

Technological innovation results from the exploitation of discoveries and inventions, and from the utilization of new materials or new processes, or the better use of those which already exist. It implies therefore a drive for new knowledge about the natural environment, and an inventive attitude of mind. The combination of these two—as we have already underlined—determines the activity of 'research and experimental development' (R & D). The term remains relevant even when the activity in question is empirical and spasmodic, as in the case of those 'inventors' whose achievements have become a supporting legend of popular novels and films.

At present, technological innovation—unlike the kinds of growth already mentioned—implies a raising of the average level of education of the labour force; in fact the greater part of technological progress can only be achieved

1. *The American Economic Review,* Vol. 57, No. 2, May 1967.

in industry, agriculture or the tertiary sector when the workers have reached a sufficient level of general education.

No doubt this is why certain authors, including E. F. Denison, have not hesitated to single out education as an autonomous factor of growth, thereby assuming that intellectual investment has, in its own right, the same kind of effect on production as the purchase of machines. However, it appears that additional investment in education would be a direct cause of productivity increase only in situations when the workers are underqualified for their task. Such a situation is far from being general. In the initial phase of industrialization, it has even been observed that the intellectual level required of the labour force has declined. The skilled urban craftsman was replaced by a machine operator, often illiterate and of rural origin.

True enough, since the end of the nineteenth century, the economy has required, in almost all its operative functions—from the farmer to the leaders of software enterprises—a higher intellectual level of education than ever before. The need for greater adaptability to change in methods of production —that is to say, the need for technological innovation—accentuates this trend. However, it is by no means proved that one cannot increase production without changing methods and techniques. That is why it is difficult to consider education as a truly autonomous factor of growth.[1]

On the other hand education is, without doubt, one of the most efficient means of propagating new technologies and methods of production. In this respect it is a powerful vehicle of technological innovation. It assures rapid dissemination of new technologies through the whole body of society, and it fosters a deep understanding by the population of the reasons for, and the advantages of, technological change. That is why we are tempted to consider the spread of education (from the point of view of the economist, naturally) as a functional need springing from development based on technological innovation, and as one of the instruments for the propagation of such innovations throughout the national economy.

But if research and experimental development are the fountain-heads of technological innovation, these activities are not necessarily carried out in each of the economies which wish to apply the results therefrom to development. New technologies can be invented in one country and applied in another. One then speaks of innovation by horizontal transfer of technology.

The ease of communications between peoples at the present day has led to increasing technological unification of the world. That is why, when one considers a single national economy, the innovation by horizontal transfer of technology tends nowadays to outclass in importance innovation by vertical transfer of technology, that is to say, innovation which results from research carried out within the national frontiers.

1. When education advances more rapidly than the modernization of economic structures, which ought to create employment for young graduates, the phenomenon of the 'brain drain' appears. Instead of stimulating growth, the development of higher education in this case tends to encourage the emigration of talented young people, and this certainly does not contribute to the growth of the economy. The non-industrialized regions of the developed countries are suffering particularly from this phenomenon.

26

GROWTH BY ORIGINAL TECHNOLOGICAL INNOVATION

However, when an economy already uses the most productive operative technologies known to mankind, it has no other means of growing except through original technological innovation. Scientific research, and the experimental development of new products and processes, are therefore the essential source of growth in the highly developed nations.

It must not be concluded from this that these countries are the only ones to evolve original technical innovations. The most advanced countries are not the first in everything; each nation, whatever its level of development, may take part in original technical innovation, at least in certain of its enterprises or laboratories, and particularly in fields where it has developed strong positions. There is, besides, the well-known fact that a potential for original technological innovation is necessary in order to get the proper benefit from other people's innovation: if the research workers do not always succeed in their own efforts, at least they know what they are looking for, and are quick to recognize it when somebody else finds it first. An industrial firm or a nation which intends to grow chiefly by horizontal transfer of technology must therefore, paradoxically, maintain busy research laboratories for the purpose of establishing intellectual contacts and exchanges that will lead to favourable transfer conditions sooner or later. These laboratories will also take part, to a certain degree, in original technological innovation.

CONCLUSION ON THE FORMS OF GROWTH

The transfer of technology and original technological innovation clearly depends on the *scientific and technological strategy* of nations and industrial enterprises. As to the advance in structural set-up and organization of the economy, which constitutes the third type of growth, it derives in some of its aspects from *systems analysis*.

This new multidisciplinary science is a part of the *scientific approach to decision-making* already mentioned in this essay. It stands where the roads of science, technology, economics and sociology meet. Here one denotes by the word 'system' the functional set-up consisting of hardware, men and information that comprises business firms, branches of industry, transportation networks, educational networks, etc. They are studied from the angle of output optimization, from the angle of effectiveness, and from the angle of the adaptation of equipment and machinery, structural relationships, and organization to the various functions of production, service, progress, etc., which the system must ensure.

In some ways, the economy of a nation—and the economy of the world—form *systems* which ought to be *optimized* anew every time that the evolution of technology or other factors modify their characteristic variables or parameters such as, for example, the threshold size below which production units are no longer functional or profitable. It is in this sense that J. K. Galbraith writes, in his operational analysis of contemporary American society, when he refers to the *industrial system*. The sum total of the scientific institutions

27

of a nation constitutes, in the same broad sense, a *sub-system* whose organization and internal and external links can be scientifically studied with the purpose of arriving at well-informed governmental decisions. The scientific approach to decision-making is, moreover, one of the functions of the science policy bodies of modern governments—an aspect which is dealt with in the third part of this essay.

Whereas some years ago the expression organizational or structural reform usually meant measures of a legal or normative character, the theory and practice of reforms has taken on a new and more concrete sense through the application of systems analysis. The approach to the problem of structures and organization has become at the same time less mechanical and more dynamic—less anatomical and more physiological. The aim nowadays is to make clear the means by which a system ensures its self-regulation and self-transformation in order to adapt itself to circumstances, somewhat as living organisms do. Growth is one of the manifestations of an organic dynamism of this kind.

For the last twenty years, numerous scientists have sought to isolate the results of the different growth factors in their attempts to identify a production function and to determine its variables and parameters.

The first attempts, made by Cobb and Douglas, only took account of the first two types of growth described above, and therefore endeavoured to account for the growth of production from a functional relationship of two input variables only: labour and capital.

In its first form, the production function assumed that both organization (or structure) of production and technology remained constant. Such a pattern clearly did not take account of the actual evolution of societies. The production function was later improved by adding a time factor to it. This attempt, however, contained the implicit hypothesis that the advances in organization of production and operative technologies are taking place through the simple march of time. Such a model did not prove very useful, because the aim of the production function is to help the decision-maker about investment, manpower policy, organizational reforms, application of new technologies, etc. It is in pursuance of this aim that one endeavours to identify the effect of the different inputs which are nowadays reckoned with in production functions.

Hence, the economists soon made an effort to measure the influence on growth of a third input, which would correspond to the increase of scientific and technological knowledge.

These efforts have considerably improved our understanding of the process of economic growth. But they have fallen short of giving policy-makers any clue as to the optimum level of a nation's investment in research or education.

This shortcoming appears to be due to a somewhat mechanistic approach to the very notion of 'input' utilized in the production function.

It now seems evident that science is not, like capital goods or human labour, a resource which one can inject at optimum rates into the process of production. This quantitative approach to scientific activity appears to be applicable only to a small number of science-based industries, where the expense for research and experimental development is on a scale comparable to other components of production, especially the investment in fixed capital. In the largest number of industrial branches and, *a fortiori,* in the economy as a whole, the expense of research and experimental development is on a scale very much smaller than that of the other resources engaged in production. It is more a catalyst than anything else. Its effect is qualitative rather than quantitative. It has not yet, therefore, the character of a *macro-economic input* whose impulse effect on the economy one can hope to measure or calculate. Proof of this may be found in the fact that the United States of America spent about 0.2 per cent of its gross national product (GNP) on research and experimental development, up to 1920, at the dawn of a decade of growth and particularly rapid industrial change, whose main driving force was nevertheless the technological and scientific revolution. This example, and many others, shows that the influence of science on society is not identical with, nor can it be reduced to, the size of the budgets for research and experimental development.

The scientific public services, the scientific approach to decision-making and management, the transfer of science and technology, etc., combine their efforts with those of research and experimental development to bring about the changes of technology and organizational structures. And so does education, which disseminates new technologies and motivates the desire for change. Several of these factors already exercise a considerable impact on economic growth, although their share of national expenditures is still small in proportion to the GNP. Further growth of the economies which operate at full employment and heavy capitalization thus appear to be brought about chiefly by a high receptivity towards structural reforms and technological change, based on a scientific attitude of mind with regard to the problems of production; that is to say, on a scientific approach which permeates the thinking of all social strata in the country.

The importance for a country of sustaining an autonomous venture in research and experimental development lies above all in the fact that this very approach to problems will be seized upon and assimilated in depth by the nation, so that the whole process becomes a definite part of its life and, if possible, is lived through from beginning to end at least in a certain number of scientific or technological areas. It seems that a nation only assimilates fully that which it would itself be capable of inventing; it should therefore devote itself to innovating activities of the same kind as the foreign discoveries and inventions whose results it wishes to assimilate.

A science-based development policy will not therefore attain its objectives by spending an arbitrarily calculated amount of funds on research and experimental development, nor by importing, on a massive scale, ready-made technological knowledge from abroad. It will aim rather at choosing and developing the best places in the substratum of the nation for the impact of scientific change. These are the places where a sufficient concentration of scientists and technologists could make the most decisive changes in the organizational structures and the operative technologies, and could thereby succeed in incorporating foreign 'know-how' into the national productive

29

system. This incorporation will have far-reaching effects if the scientific approach of these men is 'lived out' authentically and completely, that is to say if it extends from fundamental knowledge to original technical innovation and decision-making, and if it is perceived as such by the whole population. So the catalytic effect of a nation's scientific expenditure on its economic growth can be maximal, even if the areas chosen as points of impact do not cover the whole vast sphere of productive activity.

SCIENTIFIC HUMANISM—THE CONTRIBUTION MADE BY SCIENCE TO THE QUALITY OF LIFE AND SOCIETY

To free oneself from poverty by developing production is still without doubt an important objective, at the present stage, for every country in the world. We have seen that every nation has a similar attitude to this problem.

But the advanced nations can no longer find in it an exclusive or dominant motivation. These nations urgently need to give shape and substance to their other national goals. The other countries would be wise to start preparing for this too.

Beyond the physiological quantity of calories and food which are necessary to prevent man suffering from hunger—and beyond a minimal consumption of clothes and the means of heating and transport—the utility of material goods does in fact decline rapidly. The emergence in certain parts of the world, in the middle of the twentieth century, of a consumer society following on thousands of years of universal poverty, may have given rise to a somewhat childish infatuation for material goods. This enthusiasm is already generally outgrown, for the quality of individual life is clearly more important than the quantity of consumer goods.

Similarly, the quality of social life becomes more important than the *per capita* gross national product (GNP). These truths are part of the traditional wisdom of all peoples. Our generation is somewhat ashamed at having taken half a century to rediscover these fundamentals—half a century, during which a forced enthusiasm for the 'consumer society' has been the leitmotiv of so much propaganda, and the origin of a simplistic scale of values.

Can science help man to improve the quality of individual life and the quality of society, in the same way that it helps to increase the quantity of material goods produced for every hour of work? If the answer had to be in the negative, the incursion of science into the lives of men would have been a very poor show indeed.

Although our ability to give a scientific answer to the question of the conquest of happiness may be clearly less impressive than that which we have demonstrated so far in the production of electric current or motor-cars or atomic megatons, there is no factual evidence which justifies any basic pessimism on this score.

THE QUALITY OF LIFE

The quality of the individual life is capable of considerable improvement in the near future. The average improvement in health and in education is already evident.

Education is, however, only at the dawn of its qualitative development. Our techniques in this field have hardly begun to progress, and the general training which is dispensed to more and more people throughout the world is hardly different from that which was given to the élite of earlier ages in the pre-industrial societies. There has simply been added some natural science and mathematics without any noticeable modifications in either the approach to the learning process or the method of teaching. The effectiveness of such education has remained what it was before, and one is still very far, even with the most brilliant pupils, from the fullest development of their mental and physical possibilities. The less gifted boys and girls end their studies very much further yet from their optimum; the deficiencies of a mediocre educational technique being rarely compensated, in their case, by the innate receptivity and creativity of the pupil. Not to mention the countless children and adolescents who are unsuited to traditional education, or those whose development has been checked by the psychological or sociological faults of adults and who consequently suffer from inhibitions of one kind or another.

One must also draw attention to the poverty of the teaching of the arts, and to the tremendous dissemination potential of modern reproduction techniques. These techniques have made artistic creation a rare and outstanding achievement, reserved for a few dozen exceptionally gifted professionals. The works of this minority are, according to the circumstances, either endlessly reproduced or, on the contrary, placed in sanctuaries of culture, to be worshipped ritually there by the multitude. Dare we think that sharing in artistic creation could be, on the contrary, recognized as a means of personal expression necessary for every normally gifted man or woman, and that therefore this participation ought to be one of the essential aims of education?

No human being can be happy if he does not employ to the full the potential riches which he feels present within him. The refusal to give him a sufficient education, as well as the bad quality of that which is afforded him, prevents him from realizing himself. This is why this refusal is resented keenly as oppression or aggression.

In the same way, the lack of leisure of financial means during the years of childhood and youth when the powers of learning and assimilating culture are at their highest, is keenly felt by poorer people as a grave injustice. When society has at its command resources which are frequently used for artificial needs and false values, that which was merely an injustice becomes a deliberate mutilation of a whole section of the younger generation. Lastly, the indifference or the disdain with which certain societies and certain social groups receive the aspirations of the people to improve themselves by education and culture, are today resented as insults to the dignity of man.

It must be admitted that instances of oppression, injustice and insult remain innumerable, even in the most 'highly developed' societies. However, apart from cases of discrimination that are well known, these situations do not apparently fit in to any intentional plan of society on the whole in dealing with the young or with certain sections of them. It is more a matter of survivals from pre-industrial societies, in which poverty and the established laws of distribution of wealth and power made the absence of education the general rule, whereas the free development of the intellect and of culture was the lot of only very few.

By generating a need for widespread primary education, and by demanding a higher standard of education for a rapidly growing number of workers, the process of industrialization is today everywhere smashing these traditional barriers. That it should do so is indeed very much to the credit of the indus-trialization process.

But up to the moment, few nations have advanced their drive for education beyond the needs of the economy—that is, beyond the number of graduates which the production system requires.

However, in smashing the social framework of underdevelopment, and in sweeping away at the same time the shackles of poverty, industrialization also generates a widespread aspiration for an education which would know no other bounds than the possibilities of the pupil, and would be designed to make of him not just an operative needed by the economy, but a fully developed man, able to raise his life to the highest level of quality which his natural gifts can sustain.

Industrial societies have not yet produced either the teaching techniques or the educational machinery which they must have at their command to satisfy this need, and because of this they have put themselves in the danger-ous situation which leads to social strife and the conflict of the generations. This conflict, in turn, bears on the objectives which society sets itself, and the values which it professes in support of these objectives.

The accent which has for too long been placed on the increase of the pro-duction of material goods entails, as J. K. Galbraith emphasizes, an excessive value being attached not only to these goods themselves and to the possession of such goods, but also to the activity of producing them commercially. The inordinate value placed by society on the possession of motor-cars and gadgets clearly works to the disadvantage of other objectives which society could set itself, notably on the plane of education and culture. It is accom-panied by a corresponding devaluation of activities outside the circle of pro-duction and commerce, and of the people who devote themselves to them: the politicians, the writers, the Church, the artists, the students, and intellec-tuals generally. Research in exact and natural sciences or in technology, because of the direct contribution which it makes to economic growth and military power, is the only activity of the mind which finds support in the materialistic ethic.[1] It therefore benefits from a high prestige and corre-sponding advantages.

But the students, who comprise a more and more numerous social group, are conditioned—by the very values which are taught them—to react against the excessive value placed on consumer goods. They are perhaps influenced

1. In the vulgar sense and not in the philosophical definition of the word.

32

even more by the fact that most of them have hardly yet reached the paradise of the cheque book. In fact, the continuation of their studies involves a renunciation of the joys of lavish consumption, as much for the student as for his parents, even when a grant has been allotted to him. As for the minority of students who are the children of well-off families, they share in the profusion only as a result of economic dependence on their family. This dependence is felt to be less and less compatible with the dignity of a grown man or woman.

Thus all the conditions for a searching scrutiny into a society which overestimates production and consumption are found together in the student community. This no doubt explains the simultaneous unrest in this community in very different political contexts. The criticism of industrialized societies by their young people can however have a positive effect if it moves them to put their system of values back on its hinges, that is to say to adopt as an end the quality of life rather than the abundance of goods, and consequently to reshape the style and tone of society.

However, it is clear that this issue is no contradiction of values, but rather a difficulty of adaptation, for the abundance of material goods is not incompatible with the quality of life, unless it takes its place as an aim instead of serving as a means.

THE QUALITY OF SOCIETY

A similar desire for change underlies that other national goal which can perhaps best be subsumed under the expression 'quality of society'. Here also the most highly developed societies suffer from the survival of pre-industrial situations, in which a complex system of families established as hierarchies, and the compartmentalization of society, slowed down the distribution of power and so the distribution of wealth. The dialectic between order and justice dominated political life in these societies for centuries, and no one could rise above this contradiction. All the other aspects of social life were, in fact, subordinated to the vicissitudes of this struggle.

However, industrialization has the power to overcome *the antinomy order/justice,* since that antinomy derives one of its two sources in the scarcity of material goods. Thus, on the evidence of the past, and looking forward into the closing decades of the present century, it seems reasonable to predict that economic growth will eliminate this source of antinomy in a foreseeable future.

The second source of the antinomy order/justice is to be found in inequality, which the complexity of industrial society and its precariousness seem to have increased rather than diminished. At the very moment when one of the two sources of the antinomy is becoming weaker, the other seems to be growing stronger.

We have therefore for several years been the witnesses of a convergent evolution of industrialized societies towards a *technostructure* or *bureaucracy,*[1] that is to say towards a kind of society in which the central objective

---

1. *Technostructure* is the term which Galbraith uses to define the effective power which the technocrats of industry enjoy in the large capitalist societies. *Bureaucracy*

of the holders of authority is no longer merely the defence of the privileges of birth or property (these privileges may still exist, but they have become less important) but the defence of *effective powers* which ensure control. Similarly the central objective of the struggle for power has ceased to be the distribution of wealth; though this objective has kept some of its importance, it has uniformly tended to become less essential. It is found to have been replaced more and more by the demand for *participation in decision-making,* or by the still more radical objective of *self-management.*

The insistence of participation in decision-making, or on taking decisions oneself, seems in the first place to be linked with the fact that the collective blueprints of society, as they are now established, do not correspond with the aspirations of the masses and especially with those of the younger generations.

Its second cause seems to be the frustrations which spring from arbitrariness, from the abuse of power, from the brutality or indifference—or more simply from the lack of intelligence—which often characterizes the lower echelons of the 'technostructure' or the bureaucracy. These echelons are the ones through which the man in the street comes into contact with modern society at every stage of his life as a schoolboy or student, later as a workman, or tenant, or citizen, or consumer, or tourist, etc. In the pre-industrial society, where the workshop and the small business prevailed, the human contacts were certainly not always satisfactory—far from it—but they were made between two individual human beings. Bureaucratic dealings, on the other hand, are made between a man and an administrative machine.

The demand for self-management or joint control seems then to proceed on the one hand from the *desire to change* the options or the objectives which 'the establishment' or 'the system' or 'the machine' pursues, and on the other hand to alter the relations between on the one hand these organizations or social structures, and on the other the individual human beings.

Twenty or thirty years ago a large section of democratic opinion believed that it could attain these two aims by means of political power attained through democratic elections, or at least by exercising pressure and popular control on the organs of government. Clearly, this approach 'from the top' has not achieved any very sweeping success with regard to improvements on the plane of human relations.

The pressures nowadays are found much nearer to daily life. It is on the level of the firm, the organization or the university that self-management or joint control are demanded. No one can claim with certainty that this *daily-life approach* will have better results than the approach 'from the top', in preventing the alienation of the ordinary man by the social structures which make use of him for ends that he does not acknowledge, or which, in their dealings with him, exceed the limits which he is prepared to concede. Such

means rather the *de facto* power of the civil service and the political parties which control it. In combination the two form an administrative machinery which, for the most part, escapes the control of the theoretical legitimate power (the people, and, in some countries, the shareholders). This combination—the civil service plus the political parties—is therefore tempted to regard its power as an end in itself.

34

are the terms in which the problem of the quality of society will develop at the end of several decades of sustained growth, as society emerges from its obsession with the GNP and the way in which it is distributed. Our complex societies will have to discover how to preserve the efficiency and the safety of their management, while at the same time providing two guarantees to the whole of their members: the first, that of participating fully in the choice of objectives, and in the fixing of ethical limits (that is to say the full exercise of their responsibility as men); the second, liberation from the yoke—always dehumanized and often brutal—which the technostructures or bureaucracies, whenever they follow their natural bent, impose upon individuals or upon the masses.

To sum up, the antinomy, order/justice, will not be removed by economic growth. It is, however, on the way to being overcome by such growth, at least in one of its aspects. It runs the risk of being replaced gradually by the antinomy between the *efficiency of society* and the *respect due to man*.

This second antinomy will become more and more unbearable as the material goods, which at the moment are equated with the efficiency of society, become devalued as a result of their very abundance.

THE LIBERTY OF THE INDIVIDUAL AND THE LIMITS OF SCIENTIFIC ACTION

How far can science contribute to overcoming this new form of antinomy? As will be shown in Chapter Two below, one certainly ought not to demand of science that it should determine the ultimate goals of its action with the same objectivity that it applies to explaining reality or to optimizing the means in such action. The quality of life and the quality of society are subjective goals and not facts that can be submitted to cognitive judgements, such as those that science can make. The collective choices belong to the sphere of politics and not to the sphere of knowledge. But science can contribute to the choice of stages and strategies involved in achieving the ends which men have set themselves. It can help to put the objectives in a more concrete form, and to determine whether they are realistic and whether they are coherent and compatible. It also makes it possible to invent new techniques to attain the ends. In a word, it opens up prospects towards a rationality of behaviour which, to tell the truth, we are still a long way from glimpsing.

If the application of science to the solution of the problems of man and society is not more advanced, it is doubtless because behaviour is the most complex of all the biological phenomena. The attempts to discover causal laws governed by a limited number of variables have had to be abandoned in this field. Mathematical calculation and the statistical approach have certainly made it possible to uncover standard patterns of behaviour which are normally concealed by innumerable incidental factors and, as a result, to establish at least functional associations of phenomena and situations, endowed with a certain permanence, though these are not comparable with the principles and laws that govern the inanimate world. Further progress of this kind has been made, or can be envisaged. As in the other sciences, theory and observation-by-experiment progress by a succession of interactive processes.

35

The great advances made in the field of information processing (the computers) open up promising prospects to the biological, social and anthropological sciences, precisely through the possibility which they afford of handling very diverse and very numerous experimental data. But only in proportion to the progress of theory will really valuable use be made of these.

The sciences of behaviour (ethology) will however always bear the mark of the constraints which have fettered them from their very beginnings, namely: (a) experiments *'in anima nobili'* have very narrow deontological limits; (b) even when they are feasible, it is very difficult to reproduce exactly the conditions of experiment while controlling each one of the variables which determine the environment and the individuals who are being experimented upon; (c) moreover, the behaviour of man is continually influenced by what he learns about himself and others, that is to say by the most recent state of knowledge of the sciences of behaviour and of the social sciences themselves.

As an example, one can imagine that the phenomenon known to economists as the Jurglar cycle would unfold very differently if the producers and buyers were ignorant of economics. If the law of the attraction of masses, discovered by Newton, was in the same way modified by the knowledge which men have of physics, the progress of astronautics would certainly be slower than that which we have witnessed for the last ten years.

The handicap of these constraints is, then, only too apparent. However, their existence in no way constitutes a reason for doubt or despair on the question whether, or how far, the scientific approach can be applied in improving man's understanding of man (the 'sciences of man')—an improvement which is the prerequisite condition for creating a new technology of education, of management, of society and of government, all with a view to helping men attain the ultimate ends which they set themselves.

On the contrary, one must find reasons for hope in these constraints; for, if man's behaviour is modified by what he knows about himself and others —that is to say, by the progress of the life sciences and the 'sciences of man'—will he not of his own accord discard the most negative aspects of his behaviour, as his knowledge becomes broader and deeper?

Besides, if psychological phenomena were as perfectly reproducible as the conditions surrounding the fall of bodies, men could be manipulated as easily as the counterweights of a lift, or Pavlov's conditioned dogs.

The third constraint therefore preserves for man his chance to remain an acting subject, and not an object to be acted upon. This is well worth the added difficulties that have to be overcome on the road to scientific knowledge and its use in the service of humanity.

The freedom that will thus be safeguarded ought to be put to good use in singling out, for fulfilment, aims worthy of this extraordinary collective endeavour of mankind—namely the science-induced mutation of society, and the emergence through science of a new society in the world-wide sense.

2 Science and technology as factors of social mutation

SCIENCE FOR APPLICATION

INTRODUCTION: SCIENCE AND POWER

While the concept of science is ancient, the penetration of science into governmental and social practice is of recent origin. In the course of this century, scientific mentality and behaviour have progressively laid siege to areas nearer and nearer to the centre of social activity. With the appearance of military technology based on science, of organized industrial research, and then of scientific policy, science has penetrated into the corridors of power in three forms: military, economic and political.

THE UNITY OF SCIENCE AND TECHNOLOGY: THE RESEARCH PROJECT

Perhaps the best way to go about explaining the unity of science and technology is to begin by pointing out that science provides a very precise instrument of knowledge, that the practical use of this instrument brings about a profound change of technology, and that in this way science becomes the key to effective action. This was the approach adopted in Chapter 1.

The present chapter aims at a deeper analysis of what is called, in its title, *'social mutation';* that is to say a process which affects all society's values and the whole of its organizational structures. With this in view we must recall, right from the start, that the difference between science and technology tends to become blurred, and that the period—which was in any case relatively short—in which technology was essentially the application *a posteriori* of scientific discoveries, has come to an end.

Modern scientific research puts into operation complex instruments and equipment, necessitating the most advanced technologies, which depend on the solution of new theoretical problems. Moreover, technological research relies on the use of rational methods, which are exactly the same as those used in what is called 'pure' research.

One could put forward a classification of research in terms of *projects*. There are research projects which are directed towards individual consumption (for example the research for new television systems); certain projects are directed towards public service (for example, the construction of a road network or the organization of a system of preventive medicine); others have military aims (for example the construction of a device for the detection of missiles); still others aim at knowledge for its own advancement (like the installation of an astronomical observatory on an artificial satellite). What characterizes the modern age is the way in which a research project is determined and how its execution is carried out.

The definition of a research project is not only the translation into action of an affective preference, or a value judgement, or of an attitude of will. It is the result of a *process of rational decision-making,* founded on a precise appreciation of the situation, the resources, the possible strategies and on the mathematical computation of the optimal methods, paths and combinations.

The carrying out of a research project no longer implies merely the use of intelligence or the simple translation into action of ingenious or clever ideas. It relies on the application of a programme very exactly tailored to the definition of the objectives, and on the recourse, at each stage of this programme, to the most appropriate methods of analysis and experimental investigation.

THE SCIENTIFIC APPROACH

The conceptual stage and the empirical stage

What we must try to define is the process used in research, for this is the reason for the success of science, both in its powers of explanation and in its practical efficiency. As a general rule it is true to say that science is critical knowledge, that is, a kind of knowledge which is self-regulating. Since the objective of knowledge is to learn about reality, the regulation must have an effect on the relation between knowledge and reality. It must therefore rely upon a comparison between our models or representations of a reality, and this reality itself.

However, a knowledge of reality, which remained dependent on the subjectivity of the research worker, could not serve as a basis for practical application in society. The prerequisite of such application is that the concordance between reality and its representation—one might say the equation reality-representation—must be *intersubjective:* of such a nature that anyone, no matter what his subjective ideas may be, can instantly recognize its validity.

Moreover, for that mutual relationship to exist, reality must become accessible; in other words it must be 'put across'. But the only way of *putting across reality* which can be counted on to achieve the intersubjective concordance referred to, is by means of knowledge based on perception—whether such perception be direct, i.e. acquired by means of the organs of sense, or indirect, i.e. acquired with the help of instruments which extend the senses. In the past it has not proved possible to establish any lasting and truly universal intersubjective agreement on what can be imparted *a priori,* nor even on the possibility of imparting *a priori.* On the other hand it has become clear that an agreement is possible on what can be imparted *a posteriori,* that is to say on *empirical information.* This is why scientific thought accepts as a fundamental postulate that the content of knowledge is entirely based on what can be imparted empirically.

This does not mean, of course, that all observed data should be regarded as satisfactory. Agreement is not possible unless the conditions, under which knowledge is imparted, are specified in an exact manner. But the manner in which the imparting of knowledge is specified belongs to the sphere of thought: it is expressed in the form of propositions, which give rise to concepts, and concepts are a product of intellectual activity. So one is arguing in a circle: knowledge is expressed in propositions, and therefore in conceptual language: in order to test knowledge, one requires empirical data; but these data themselves can only be apprehended, isolated and classified by the help of a perfected conceptualization. So it is clear that the scientific process requires two stages, a conceptual stage and an empirical stage, and that these two stages are indissolubly linked.

The link between the conceptual stage and the empirical stage

One must, however, define exactly the nature of the link between these two stages, and that makes it necessary to establish from the start the distinction between two categories of concepts: *theoretical concepts* and *empirical concepts.* The latter denote qualities directly distinguishable in observation, and are used to describe the actual situation. Theoretical concepts, on the other hand, have no direct relation with experience. They are the components of a coherent and logical structure in whose terms we can form a representation or model of the sector of reality which concerns us.

Of course theoretical concepts are only useful if it is possible to pass from the language in which they can be framed to the language in which one describes the actual situation.[1]

Therefore every theoretical construction is accompanied by rules of interpretation (for example a theory of the methods of measurements involved), which make it possible to translate at least some of the propositions it contains into propositions describing some aspects of perceptible reality. We thus find ourselves confronted with two kinds of language: a language of theory (accompanied by its own rules of interpretation, which are themselves of a theoretical nature) and an empirical language. It is on the level of the language of theory that the scientific models of the world are created, but in themselves they do not constitute the whole of science.

Within the framework of theoretical language it is possible to formulate propositions which play the role of hypothetical representations of the subject-matter under study. These hypotheses must be put to the test by a

1. For example the theoretical concept of the field of gravity (*champ gravifique*) is only useful because one can pass from it to the empirical concept of acceleration which is directly comprehensible and measurable.

comparison between their conclusions and that which can be actually observed. In so far as the hypothesis stands up to the tests to which it is submitted, it is considered to be confirmed; if it is contradicted by certain facts, it is unsound and ought to be rejected.

The essence of the scientific approach consists, therefore, of two operations; the *invention of hypotheses,* and the *testing of these hypotheses* by confronting them with the actual situation. But putting a hypothesis to the test often involves seeking out facts which do not present themselves spontaneously in nature, and whose presence therefore has to be elicited by appropriate operations. This approach is called experiment.

However, experimental operations themselves presuppose, at least as a temporary expedient, the recourse to certain theories and therefore to certain hypotheses. Thus the use of an electrical apparatus presupposes the theory of electricity. The testing of a theory is only possible to the extent that one provisionally accepts the validity of a certain number of other theories. It is therefore never possible to isolate the theoretic stage completely from the empirical stage.

On the other hand, although theory (that is, the hypotheses which we formulate with the help of our theoretical concepts) may have the function of providing us with a representation of reality, it is often by complicated changes that one passes from theory to the empirical data and vice versa. And though it may be that the hypothesis is suggested by certain facts, nevertheless once it is formulated, the hypothesis in some ways outstrips the facts: it goes further than the facts which have suggested it, it forms a kind of question concerning other possible facts. When one puts it to the test, one's efforts are concentrated on bringing out into the open the facts which it is assumed to imply. It is therefore the theory itself which, by being put to the test, guides the empirical investigation.

If one had to be content with the accumulation of situational data, as they happened to present themselves, one would have great difficulty in distinguishing the facts which were really significant from the others; moreover a great deal of time would be lost in cataloguing facts totally devoid of interest. An intelligent investigation must be guided by an idea. In other words, research must give the answers to questions: only in this way can the observed facts be appreciated in their true meaning. Therefore the hypothesis —that is the theory—has a logical priority over the empirical data. From the historical point of view, the hypothesis presupposes the knowledge of certain facts; but from the logical point of view, and from the point of view of the strategy of research, the hypothesis precedes the facts and phenomena discovered by experiment, and controls the way they are brought about.

These schematic details about the relations between the conceptual stage and the empirical stage of science will already have made it clear that science, far from being a contemplative knowledge of reality, is in the strongest sense of the term a practical application of knowledge.

In fact, a hypothesis is a tentative representation of reality, its true role being not to provide us with a likeness of reality, but to give direction to research, and so to guide finally to the formulation of other hypotheses, more fruitful and more rewarding. A hypothesis is never anything but a temporary representation. The life of science is in the incessant flow or movement from the hypothesis to the putting to the test, and from that to the criticism of existing hypotheses and the formulation of new ones.

THE EFFECTIVENESS OF THE SCIENTIFIC APPROACH

The operative aspect of theory

It is important to emphasize, in the context which concerns us, that not only is science a practical application of knowledge, but that it should be an efficient one, as much in the theoretical stage of its process as in the empirical stage. This is why the language used to convey theoretical concepts has been formalized. It is true that, as yet, we do not have at our command a complete formalization, which would make it possible to avoid using natural language altogether; we continue to make use of the latter, if only to explain how to use the formalized parts of language. But to the extent that the language of theory becomes established, it is possible, in principle, to stretch formalization as far as seems desirable. The formalized language of theory includes the mathematical language. But it also contains terms which make it possible to express non-mathematical propositions in a formalized manner, playing the role of axioms in reference to the body of propositions that one wishes to produce (as with the axioms of mechanics). As has already been indicated, these terms have to have a definite meaning, but the rules of meaning themselves can be formalized, thanks to the methods of modern semantics.

The first complete theories of physics used, in their strictly mathematical part, the language of the infinitesimal calculus, in accordance with the methodological postulate which consists in regarding the properties studied, and their variations, as showing no discontinuity. So quite naturally one is induced to represent them by real numbers and to represent their interrelations by continuous functions of real numbers.

But in the majority of more modern theories the language has been considerably widened, and it is the tool of algebra which appears today as the favoured instrument of representation. Algebra, indeed, makes clearer than the infinitesimal calculus what the fundamental characteristic of formal language really is—namely that it is an operative language. Modern studies in logic have emphasized this characteristic of formal language and have discovered in algebra the ideal tool for formal analysis; algebra, in fact, is the science of operations as such and of their properties. That which, from the point of view of algebra, characterizes a domain is not the particular nature of the objects which form part of it, but the nature of the operations which can be carried out on these objects.[1]

1. Thus the concept of a group (such as is used in modern algebra) is independent of the substratum on which the group is defined. Rational numbers form a group with respect to the operation of multiplication. But the same applies as regards the rotations of the sphere, with respect to the operation of combining two rotations with each other. This last example shows us moreover that the elements of a group can themselves be operations.

As is made apparent by the study of *combinatory logic,* which is a veritable 'formalization of formalization', the method which makes it possible to bring language back to a purely operative use is precisely—formalization.

Whereas, in natural language, the meaning of an expression is shown by its reference to extra-linguistic elements, which belong to the physical world or to the sphere of behaviour, in formal language the meaning of an expression is shown by the way in which it can be employed. Its intelligibility is therefore of an operative nature.

Let us consider for example the operator of inversion, *I*, which one defines as:

$$I \quad ab = ba.$$

The use of the term 'inversion' clearly recalls a meaning which the word possessed before it was formalized—a meaning borrowed from the manipulation of objects in everyday life. But in actual fact its use here is purely extrinsic. One could simply speak of the operator *I*. The real way to understand the meaning of *I* is to refer to its formal definition and carry out the prescribed procedure. It would appear that the extension of formal vocabulary is in this way narrowly limited to operations of the type of those which one can establish by means of *algorithmic devices.* These devices are characterized by the fact that they can be put into action; if need be, they can consequently be represented by means of material structures, for example by means of computers. But nothing prevents operations of a non-constructive character—that is to say those which cannot be completed in a finite number of stages—from being just as well represented in formal terms. It is true that, in this case, the results obtained do not have the same status, with regard to their effectiveness, as those obtained by algorithmic procedures, but they belong equally to the sphere of operative processes.

The virtue of such a language is that it is in all respects a practice: it substitutes—for comprehension by intuition, or by analogy or by sympathy—comprehension by logical process, by action, by carrying out what is meant. One could call such a mode of comprehension the *intelligibility conveyed by calculus.*[1]

Calculus can give us a key to the efficient comprehension of reality, to the extent that the nature of this reality itself is essentially operative. True, this cannot be stated as an absolute thesis: but the very basis on which the use of formalized theoretical language is grounded is the epistemological postulate according to which the decisive aspects of reality are of an operative nature.

It is this postulate which has been successfully used since the earliest advances in modern physics. In this way the calculation of a parabola explains to us the behaviour of a cannon ball. What it is necessary to understand in this case is not, after all, the nature of gravity, but the manner in which gravity operates. And when one succeeds in finding a mathematical

---

1. For this purpose the word 'calculus' is to be understood in the sense—at once very precise, and very wide—attached to it by the science of formalism: that is to say, as embracing not only numerical calculus or traditional algebraical calculus, but every kind of formal manipulation tied to conditions of effectiveness.

function which expresses its mode of action (in this case the function which characterizes the parabola) one can analyse the properties of gravity in examining the properties of the function. No doubt there is an element of arbitrary simplification about representing reality in this way: no doubt there is more to reality than emerges when one chops it up piecemeal in this way. However, this kind of chopping up, and the corresponding postulate relative to the operative nature of physical reality, are justified by the fact that they prove themselves fruitful by leading to a representation which makes it possible to work on nature.

Operative character of experiment

It is self-evident that the empirical stage of science is itself also essentially operative. As has already been emphasized, the empirical investigation of science in no way consists of passively cataloguing the data, but in putting hypotheses to the test. If the hypothesis is of an operative nature, its testing must inevitably have the same character. An experiment involves making a phenomenon appear in conditions which are controlled as completely as possible and which one can vary in a systematic manner. It involves putting a question to nature, and this question always takes the following form: what will nature's reaction be if such and such constraints are put on her, which do not necessarily occur 'naturally', but which can be artificially created, by making use of phenomena already well known and satisfactorily under control? Theoretical reasoning describes in advance the expected effect. The experiment must show whether—in the conditions used in the reasoning—this effect is actually produced; that is to say, whether nature operates in the same way as the theoretical scheme or model which is being used.[1]

The overlap between the scientific approach and social behaviour in historical times

The penetration of science into the most varied and the most important forms of *social behaviour* is not, therefore, a purely accidental fact, to be ascribed to the enlightened will of certain individuals or certain groups or to a chance encounter between interests and methods. It is a fact which is historically necessary in the sense that it expresses a *law of transcendental nature*.

Science is intrinsically operative. It is, indeed, the only approach which is truly rational. Such an approach cannot avoid spreading progressively to every sphere of existence. In the beginning it devoted itself to the problems which it had inherited from ancient learning and which were of a rather speculative nature, as for example the problems which had been posed by ancient astronomy. It then spread to spheres which were no longer in the

---

1. In this way the reasoning which can be based on the hypotheses of wave mechanics makes us see that a beam of electrons diffracted through a crystal lattice ought to give patterns of interference. The experiment involves submitting a beam of electrons to this kind of condition effectively. In answering 'yes' in this case to the question put, the electrons perform on their own account an operation which undoubtedly is not identical to that of the calculus, but which one must assume is in any case analogous to it.

category of learning as such, but in the category of action. We can no longer, today, trace a clear line of demarcation between these two categories.

Correspondingly, social behaviour, to the extent that it has broken loose from magical, mythical or ideological incantations, needs to be rational and must, to this very extent, necessarily become scientific. So in the final analysis, it proves impossible to continue to divorce social behaviour from the scientific approach.

This historical overlap has naturally a profound significance: if it adopts the character of necessity, it does so because it is itself only a manifestation of mankind's fundamental destiny, *the progressive emergence of reason.* We cannot define reason *a priori,* and neither can we define its *'telos'* except in a purely formal manner. It can only be said to be that part of itself which our actions reveal to us: the *conquest of the concept of rationality* is inseparable from the *conquest of the methods* in which it is embodied.

But we know that these methods will not be perfected in a single day, and that they are not fixed once and for all. Their systematic development produces new problems, the study of which brings to light more accurate and more powerful methods. In this way the development of mathematical methods has progressively led to a more and more exact intellectual grasp of the nature of formalism and its powers. However, the development of the science of formalisms has, in its turn, led to the discovery of foundational problems which are far from being resolved, but whose exploration already makes it possible to get a glimpse of new depths of the formal method itself. So the demands which appear at the level of the methods themselves determine, from within, the *evolution of rationality.*

The same evolution is taking place in the scientific approach and in social behaviour. Just as scientific reasoning improves little by little, as a result of the way the methodological activity reflects light back upon itself, so government and social practice, as they develop, bring about an emerging *rationality of behaviour* leading to a more and more precise mastery of social behaviour over itself.

THE SCIENTIFIC APPROACH TO SCIENCE

This being so, it is natural that science should apply to its own processes the method which it has succeeded so well in putting to use in the exploration of nature. What began as a process of investigation of nature such as it occurs to the scientist from outside, now becomes the universal means of action. And as research itself is action, the idea of a rational conduct of the scientific process itself, that is to say a scientific organization of research, necessarily forces itself upon us.

Of course the scientific planning of decisions cannot adopt the traditional principles developed by the natural sciences just as they are. The first of these laws is that of *mechanical determinism,* whose usefulness in the exploration of physical phenomena, and especially in celestial mechanics, assured the triumph of the scientific approach in the seventeenth and eighteenth centuries. This principle allowed simultaneously the determinateness of the values assumed by the variables[1] and the symmetry of prediction and retrodiction.[2] The probabilistic approach, coming rather later, whose usefulness has been proved by the study of random phenomena, abandons the determinateness of the values assumed by variables, but keeps the symmetry of prediction and retrodiction.

In the domain of action, the scientific approach gives us the means of predicting the results which one can expect from the fixed initial choices, sometimes by deterministic methods and sometimes by probabilistic methods. Starting from there, it takes stock of the decisions possible within the margin of choice which actually exists in the conditions of the problem under consideration, and compares the results one must expect to follow each of these decisions. But it is not usually possible to infer the initial choices from the results, because the same final situation could be obtained from different choices.

Since scientific practice has itself become an object of science, it is possible to construct instruments for assessing the values of the different strategies open to research and action.

In this way the scientific preparation of decisions for governmental and social behaviour leads to new principles of causality whose justification will be found, like the justification of the principles which preceded them, in their ability to produce results.

THE CHOICE OF OBJECTIVES AND THE PROBLEM OF THE NORMS

INTRODUCTION: THE QUESTION OF THE GOALS OF RATIONAL CONDUCT

The formal definition of the optimal action is only the expression of a demand for efficiency which is a part of the very concept of rationality. It cannot therefore be divorced from rational conduct. If there is a scientific approach to science, it must necessarily, according to the very concept of science, be subjected to conditions of efficiency that are as stringent as possible.

What matters in all practice, much more than the approach or methods used, is the results. The rationality of the process is only fruitful in so far as it makes it possible to anticipate the results and change direction towards the goals which are set. Every action is relative to a project, and an approach or a method only have meaning to the extent that they are deemed to lead to an objective.

An initial choice therefore must be made: that of the ultimate goal of the action.

1. That is to say that at each moment each variable possesses a unique and well-defined value.
2. If the state of the world or the system at the moment t is granted as known, one can calculate its state at the moment $t + dt$. Reciprocally, from the state of the world or the system at the moment $t + dt$, one can deduce its state at the moment t.

But there is still a second choice to be made in the category of the goals: in fact the same objective can be attained in many different ways, which are not necessarily equal in respect of such considerations as cost, delay, risk, etc. The rational procedure, therefore, will be not only the one that is appropriate to reach its ultimate goal, but also the one that represents the best arrangement with regard to the subsidiary ends which will have been chosen as norms or criteria of optimization.

So in order to decide the most rational method of getting from Paris to Tokyo, one must make up one's mind from the start whether the journey has to be the quickest, or the cheapest, or the one which presents the minimum of risks of accident, or the minimum of political risks, or even the maximum of opportunities for business contacts for the traveller.

The idea of *optimization,* which plays such a fundamental role in the scientific preparation of decisions, only has meaning in relation to a norm laid down *a priori,* often in quantitative terms, in the shape of a principle of maximization of social utility or cost effectiveness.

Rational action is therefore finalized in two ways at two different levels, the second controlling the first. So far as its content is concerned, it is finalized by the objectives which it sets itself; so far as its form is concerned, it is finalized by the norms which it lays down for itself.

PROBLEM OF THE RATIONAL CHOICE OF OBJECTIVES

The choice of subsidiary ends, from which the norms of optimization are derived, is nearly always clear enough: the time, the money, the men available are the scanty resources, and, according to the circumstances, one or an other is the most compelling motive. It is perhaps on the question of the determination of the acceptable risk that the principal difficulties are found.

But it is the question of the origin of the ultimate goals, the goals or objectives themselves, which is undoubtedly the most thorny problem. Will it be enough to say that the determination of these objectives is a matter of free will, and that it derives from a political choice (if it is the decision of a political movement) or from affective preferences (if it is the decision of an independent scientist) or again from the profit motive (if it is the decision of an industrial enterprise)? Certainly not, for one would immediately have to ask oneself how the political choice or the profit motive can be justified, or how the affective preferences can be made explicit and eventually explained rationally. It is clear that the determination of the ultimate goals is not purely arbitrary, and that it obeys norms which are not the formal norms of efficiency which we have just been considering.

One can indeed imagine an action which would be rational in its methods and completely irrational in its objectives. Nevertheless, if there is actually an evolution of rationality in action, and if it becomes further and further detached, as we have suggested, from representations of a mythical or ideological character, or from the brutal affirmation of the personal or collective ego, this seems unlikely: rationality is not like a water-tap, to turned on or off at will, so that one could hardly describe as rational an absurd action rationally conducted. The question to be asked is how to know what kind of rationality the fixing of ultimate goals brings to the fore. Let us pass in review several possible roads that might be taken.

HYPOTHESIS OF AN EMPIRICAL CHOICE OF ULTIMATE GOALS

One could from the start consider objectives as hypotheses. The rational determination of objectives would then consist in putting to the test several objectives, either successively or simultaneously, and keeping that which stands up best to the test. But to apply such a procedure, one would need to have at one's command a criterion which would allow the formation of a judgement on the acceptability of an objective. When the test is applied to a physical phenomenon, the criterion is available: it is *the criterion of whether the reality conforms with the hypothesis*. When it is applied to the action, the criterion itself poses a problem: to judge the result of a line of action, one must be able to compare it with a result that is considered desirable. And that calls for a definition of 'desirable', which in turn calls for a *criterion of desirability* by which to judge the desirability of the desired result.

It is clearly useless to resort to such ideas as those of social utility or satisfaction because such ideas are either purely formal and can be used to justify anything, or else they are purely subjective and so do not admit of intersubjective agreement.

The direct appeal to the empirical method (the method of judging hypotheses) only puts the problem of the goals further off. It only makes it possible to choose the objectives of action if the criteria of desirability are given.

HYPOTHESIS OF SCIENCE AS A PRODUCER OF ITS OWN GOALS

It must here be observed that science, as it develops, sets itself new objectives for research, as much at the theoretical level as at the empirical level. At the theoretical level, the difficulties encountered in the use of existing representations of reality pose precise theoretical problems, which it is necessary to resolve if one wishes to be able to continue. Thus, the difficulties involved in the electrodynamics of moving bodies have compelled physics to upset the theoretical structure of traditional mechanics.

On the other hand, the use of technology, which stems from science, creates new problems to which one can only hope to find an answer by a more rigorous application of the scientific approach. For example, modern industry and military agencies introduce into the biosphere disequilibria which upset the natural equilibria and regulation mechanisms, and which have to be compensated by new actions based on science and technology. In this connexion one may cite the phenomenon of over-population, which is the consequence of the spread of the use of new medical knowledge. The disequilibrium thus produced can be corrected, thanks to a radical transformation of the methods of food production, e.g., in the shape of the artificial

47

synthesis of proteins, or by a modification in the fertility rate of married couples, or more probably by both at the same time.

Science, once its processes have been set in motion, thus to a certain extent spontaneously produces its own goals.

At first its goals were governed by objective needs or by pre-scientific ideas or concerns, of a philosophical or ideological nature. But once the process has been well primed, it expands by its own resources: it inevitably produces needs which cannot be met except by new developments; this, in its turn, will bring new needs into being, and so on. In this respect one could speak of an *endogenous development of science,* and, in a more general way, of rational practice.

But actually this simply means that once the scientific approach has been accepted, certain norms are adopted as being self-evident. If one's attention is directed to theoretical problems, one discovers that there is no need for any reshaping of the theory except in relation to a *norm of the effectiveness of the theory* itself, that is to say, in relation to a guiding principle of an epistemological nature. This principle usually remains implicit, and serves as a guideline for theoretical thinking.

In the same way, in the practical sphere, the actions undertaken to correct the disequilibria that spring from preceding actions, either find themselves subordinated to the ends of the latter, or else call upon goals which are thought of as self-evident, but which are not, in any case, imposed only by the existence of the disequilibria. (For example, if the development of certain industries causes pollution of the air, the general opinion is that this must be fought against, because it is dangerous to the health of the population. The objective of health, as such, does not originate in the sole fact of the pollution of the air.)

The production by science of its own goals, and, in more general terms, the endogenous development of rational action, does not therefore offer a solution to the problem of the ultimate goals of action. No doubt it contributes to the accurate definition of certain objectives, but it can only do this in relation to goals which must have their origin elsewhere; thus its intervention is still only subsidiary.

HYPOTHESIS OF BIOLOGICAL OR 'NATURAL' GOALS

In human practice, of which science is a part, the goal sometimes allows itself to be reduced to a need for survival.

It would be tempting to reduce such a norm to biological considerations or, to be more precise, to the scientific theory according to which life always strains to invent more appropriate forms of adaptation. In this way it gives rise to a norm of optimization: the increase of the chances of survival, if not at the level of the individual, at least at the level of the species.[1]

1. It may also be observed that normative principles of this kind are common in the natural sciences. Thus there are found in physics, under the class of principles of *'extremum'*, controlling norms which produce a pseudo-goal (which is moreover purely immanent) at the level of mechanical phenomena. Behaviour conditioned

Be that as it may, the statement of a natural norm must not be confused with the adoption of guidelines for action. If we decide to make it our aim to increase the chances of survival of the human species, we thereby transform an external normative principle into an internal normative principle, a principle of natural self-regulation to a principle of rational conduct. However, there is no logical transition from the one to the other.

The natural laws of evolution (to the extent that such laws exist) are in no way a restraint for our conduct. If we continue to acquire control over evolution it means that, in this very measure, we acquire the power to counteract its laws. No doubt natural laws can suggest to us a line of conduct. But this would then be in the form of a principle, which could be formulated as follows: 'We ought to act in such a way that the standards of our action agree with what we can know of the natural laws governing the procedures which are of significance to us'. Such a principle does not in any way derive from the scientific laws on which it claims to rest, nor on the scientific approach itself: between a natural law and a standard of action there is a basic difference in kind, a radical heterogeneity. As has been fully shown, a normative judgement cannot be reduced to a judgement of fact. And when propositions concerned with goals simply state the existence of a process of optimization in certain phenomena, they are only judgements of fact, cognitive judgements: they have no effect on human action.

The use of a norm of this type implies a certain conception of human conduct: it represents men's actions, albeit in a way that is not thematic, as a simple extension of natural processes (physical or biological) and sets men the objective either of reproducing these processes, using them as a model, or else of assisting them by intervening at the points where the processes find themselves thwarted by reason of the interference of other phenomena considered as parasitical. The first case is illustrated by the old idea of art, which 'imitates nature', and the second by a certain idea of medicine as a means of combating communicable diseases. But what all experience of rational practice shows us is precisely that this equating of action with a natural process is highly debatable. When action models itself on a certain natural process, it decides of itself to be linked in this way, without there being any compulsion by a law of nature. And on the other hand, rational practice itself modifies the situational data, both around man and within man himself, in such a way that action becomes more and more capable of controlling the initial conditions: it is true that the processes continue to obey the laws of nature, but the initial conditions are less and less determined by natural processes and more and more by the human acts which preceded them. And at the same time, action is always opening up for itself new possibilities. For example, from the moment when nuclear energy has been successfully mastered, the manner in which that energy will be used is no longer in any way determined by 'nature'. The solution which would consist in looking for practical norms in judgements of fact, that is to say in the

by regulating norms obviously presents an analogy with that which results from an ultimate goal.

observation of the natural course of events, becomes therefore more and more illusory.

THE DECIDING OF GOALS A PRIORI

The two first hypotheses considered (the empirical selection of goals, and rational practice deciding its own goals) did not solve the problem of the 'determination' of goals. Both hypotheses appealed to a test of efficiency, or to conformity to criteria, or to goals settled beforehand; in this way they caused the revival of the problem which they were intended to solve. The third hypothesis, of the 'naturalist' type, tried to make an appeal to objective data, provided by the natural sciences. Such a process can be qualified as being of an *a prioristic* nature. However, as has been seen, it does not appear satisfactory. This seems to suggest that the determination of the goals of action should be looked for in reasons which have a certain stamp or 'apriority' in relation to the natural situational data. But it is important to make quite clear what sort of 'apriority' we are discussing.

If one turns to one of the traditional doctrines of *a priori* one comes up against some well-known difficulties. These doctrines argue that there are data independent of empirical experience, which belong to the sphere of subjectivity, and that it is possible to prove their existence by reflective analysis. They deal, for example, with innate ideas, or else with pure forms determining the structure of the subject, or with affective elements or with values (comprehended by intuition). The fundamental objection that can be made to methods which postulate *a priori* data, is that they have never, at least up to the present time, made possible a real intersubjective harmony. It is true that the philosophers who have worked out the theories of *a priori* are not isolated philosophers; they have had disciples and still have their partisans. But besides the fact that other philosophers purely and simply deny the existence of *a priori* data, the methods that their defenders put forward do not appear—by contrast with 'empirico-formal' methods—to be of a nature to make a general agreement possible, or even a provisional one. The fact is that the methods which postulate *a priori* data are not of an operative nature. We have seen that if empirical and formal methods make agreement possible, it is because they have the nature of a practice. Agreement is not identity of outlook, or sharing of evidence, but a tendency towards combination in action. Moreover, the theories of *a priori* which have been put forward in the past have always remained at levels of extreme formal generality, and have shown themselves therefore not at all productive in the effective definition of ideals of conduct.

CREATION OF THE NORMS DURING ACTION

It does not seem that one can dispense entirely with a certain measure of 'apriority', in the sense that one cannot rely entirely on propositions of fact. But equally one cannot adopt a method which would make it sufficient to appeal to a subjective *a priori,* assumed to be directly comprehensible (by intuition or reasoning power). Undoubtedly one must distinguish two stages in the use of the norms: the stage of the statement of the norms, and the stage of their justification.

The weakness of the doctrines of *a priori* is in fact that they treat human reality as given to itself once for all and accessible—at least in principle—to reasoning (independently of any application) not only in its sheer power to think and act, but in its content. From this point of view, action would only be the deployment of this content, which was present in its own right, in a non-empirical guise (for example in the shape of intellectual intuition or of a pure perception of values).

However, the experience of history in all its forms, and particularly the history of scientific thought, teaches us that human reality is a perpetual self-creation, which does not cease to improve on itself and to invent a future which is always unforeseen. It is not *a priori,* in the solitude of pure reflection, that it can reveal itself, but only after the event, when it is already engaged in its creative activity. Incomprehensible to itself, it can only make itself understood through what it reveals of itself, that is to say, by its works. Action is self-revelatory, for it is only in putting itself to the test that it discloses its powers. Not in the sense in which the actor reveals hidden potentialities, but in the much stronger sense that it decides its own evolution by conditioning itself in an irreversible manner. Action is not only a guide to knowledge; it is, in very truth, a creation.

This is perfectly illustrated by scientific practice. Scientific thought cannot be envisaged as the simple putting into operation of a body of concepts available beforehand and unchangeable—if only men were conscious of them—as if the virtue of an empirical inquiry was only to make thinking progressively more conscious of its means of intelligibility.

Research creates problems in making use quite naturally of concepts already available, but without being certain that these concepts will be able to succeed in unravelling the situation. And in the most conclusive instances, it actually appears that the existing conceptual apparatus is inadequate, and that a restructuring, more or less fundamental, is necessary. New concepts are therefore created, and a new structure of intelligibility appears. Of course, the new conceptual apparatus must show itself capable of including in itself the virtues of the out-of-date structures, but one cannot say that it was implicit in them in a potential state, nor that it was in some way fore-shadowed.

Every *conceptual creation* thus indicates a stage of severance: it is truly the birth of a new norm of thought. It is true that such a creation does not happen by accident. It is in some way conditioned by the kind of problem from which it springs; moreover it only justifies itself by its capacity to regain what is already acquired and open up new roads for research.

The same is true of action, that is to say the behaviour by which the human being tests himself individually or collectively, and of his own free will risks himself, stakes his future and proves himself by modifying his environment. In the same way that thought forges its concepts for itself, on the very situations in which it is involved, so action forges for itself its

51

standards in the very circumstances of its exercise. In the same way that thought only establishes itself as such in conceptual creation, so action only plays its part as such in *axiological creation*.

But in the same way that the new concept becomes, so to speak, an essential part of thought by virtue of a kind of internal logic which excludes anything arbitrary, so the new norm becomes an essential part of the action by virtue of a kind of autonomous development which prohibits anything accidental.

In conceptual creation there are severances, but at the same time alliances between systems: a new system is not totally divorced from previous systems, it takes to itself anything valuable which they included, by reworking in a radical manner the foundations on which they were based.

In the same way there is not a total lack of continuity between the normative systems. Even if there is a severance, it is always clear after the event that the birth of a new norm is a deepening of the demands brought about by the old norms.

On one side there is an evolution, with all the unexpectedness and apparent hazards that it entails. On the other side there are links which can be construed retrospectively as the manifestations of a necessity. Hence it is not *a priori*, in pure reflection, that creation takes place, but in a concrete historical problem. An act worthy of the name is never simply the application to a given situation of a pre-existing norm, which it would be enough to invoke at the required moment. The act is a creation, at the same time, of a practical solution to a given problem, and of the norm in accordance with which this solution is produced.

There is therefore a *life of the norms*. When, apparently, the same norm is invoked at different moments, this is done according to the many special modalities which diversify and enrich its meaning. One could say that man creates himself in action, not only in the content of his culture and his mental equipment, but also in his human character, that is to say in the quality with which he means to endow his life.

When the exterior compulsions are very strong, the *field of axiological creation* is very restricted, and is limited to the most pressing human relations. When the possibilities of intervention in the exterior or interior sphere increase, the field of axiological creation expands at the same time. It is this which distinguishes the modern age from the eras which preceded scientific and technological change. Man now establishes more and more freely his relations with his environment and the manner in which he intends to arrange his life.

One could get the impression that the essentials of the conduct of the human race had been dictated by biological laws, and that its first aim was to survive and to extend its control over its surroundings, to ensure its safety more thoroughly. But this statement only covers the most superficial aspect of the subject with which we are dealing, and also the least constant, since for the future the survival of the species is no longer directly threatened by its natural environment.

What is really important for man is the way in which he gives values to life, the way in which he transforms it and gives it meaning. Life for him is never a simple set of biological events. It is that which emerges by making use of its biological environment, and which seeks to expand: it is the kingdom of the spoken word, that is to say understanding, communication, presence, recognition.

A very archaic term could perhaps serve to express what shows itself in life as a demand that enhances life and at the same time recaptures it: the term 'harmony'. In an ordinary sense harmony means an equilibrium, a happy functioning, a successful adjustment of life to its conditioning, a judicious arrangement of its external environment to suit its aims. In a deeper sense—which incidentally also includes the ordinary sense—it denotes a way of life in which everything is well attuned, in which a man's life is in accord with his speech, and conversely the human personality is conscious of being in intimate communion with the world and feels itself carried away by a ceaseless flow which imbues it with a feeling of joy.

It is certainly this need of harmony, which comes from the depths of life and draws it towards its fulfilment, which is at work in the activity of consciousness which lays down the norms. No normative system has ever been completed. The axiological effort is always at work. What it achieves is always an inadequate approximation. But it is given its objective by a central need which gradually becomes more specific as the field of possibilities widens.

JUSTIFICATION OF THE NORMS

Thus one comes to have a glimpse of what could constitute the stage of justification.

In the sphere of scientific thought, there is a double justification: on the one hand scientific thought must show the internal coherence of the theoretical propositions; and on the other hand it must show their conformity with the available empirical data. In the sphere of action this double form of justification will again be found: the successful formation of a coherence and conformity with the material conditions of action. The justification does not appeal to pre-existing patterns which it would be possible to comprehend without regard to the action itself. And in this sense it does not derive from a method that is purely *a priori.*

The *condition of coherence* is concerned with the relation of the proposed norm to the other norms involved, whether we are dealing with previous norms, considered as always valid, or those proposed simultaneously in the other sectors of action. This condition is of a formal nature.

The other condition—which one could call *empirical,* since it is concerned with the relations of the norm and material action—consists in demanding that the norm should render the action at the same time efficient, self-consistent and productive. Efficient: that is, truly capable of responding to the given situation in its particular nature. Self-consistent: that is, capable of conforming in its response with the dynamism which drives it, with the very demand of normativity in itself, or again with that desire for harmony which was discussed above. Productive: that is, capable of extending this dynamism to the future, of opening up new spheres for its creativity, of giving deep satisfaction to its internal demands.

Action is fundamentally a creative dynamism, whose inner law is a continuous urge towards improvement and expansion. This law represents an infinite demand. By its very nature it enhances even its data, and it cannot be controlled by the situational data. The justification of the orientations of action cannot therefore consist in a comparison of the action with some reality exterior to it (though it ought to make allowances for empirical constraints), but only in a comparison of the action with itself. In other words, it is only open to a method of reflection.

However, this reflection is not an attempt to make visible what has been determined innately. It is engaged in an effort to keep under control the internal demands which determine it, during the very exercise of the action and in face of the challenges which are presented to it, and with which it presents itself.

It must be recognized that such a method of justification cannot be considered identical with that which takes place in the scientific sphere. It is not based on empirical verification in the strict sense of the word, but on an assessment of the action by itself; it is expressed in judgements which belong more to the aesthetic category than to the cognitive, and, through this very fact, it does not lend itself as easily as the scientific process does, to a process of intersubjective agreement.

However, this kind of thought contains nothing that is arbitrary or irrational. It can create for itself its concepts, and express itself in propositions offered for debate. The justification of the norms of action, although it does not derive from scientific reasoning, is not therefore divorced from reason. Beside *empirico-formal reasoning,* developed by scientific thought, there is room for a *reflective use of reason* entrusted with the task of controlling intelligently the normative orientations of action.

THE IDEA OF SCIENCE POLICY: EFFICIENCY AND RESPONSIBILITY

The necessity for a rational control of evolution

Science, by virtue of its efficiency, today opens up huge fields for human action. It will open up still other fields, as yet now unsuspected. By this very fact, it considerably enhances the indeterminate quality of action, and presents action with an increasing number of choices, some of which are bound to have a decisive effect for the future of the human race. Hitherto, scientific and technological progress has been the result of very haphazard decisions, with virtually no guiding principles, and made for the most part in complete unawareness of their true significance. The decisions of governmental and social practice were moreover based, at least to a large extent, on the extremely unambitious aims of subsistence, or of domination by certain tribes or nations. Our age is working out methods of organization and

54

forecast-making which ought to make it possible for us to take decisions that are better co-ordinated, more responsible for their short- and long-term consequences, increasingly effective and at the same time increasingly far-sighted.

We have therefore embarked on a process of mutation which we have ourselves set in motion, and we can have no alternative but to guide it right to its conclusion.

At the present stage in fact, the mutation remains incomplete; as a result of the action producing its own goals, the consequences of our haphazard decisions could still produce irreversible and unforeseen evolutions, and thus drive us into disaster. We could find ourselves, at a given moment, in the situation of a chess player who abruptly finds himself in a position of checkmate without having realized the significance of the moves which brought him to this position. It is therefore extremely urgent for us to retain rational control of our evolution in our own hands. This control ought to be twofold. It ought to make us conscious of the consequences of our choices even though they are remote, and it ought at the same time to make us conscious of what we truly desire.

Science policy, a rational combination of the rational and the reasonable

Science policy ought to be more than a scientific use of science. There is no policy, in the full sense of the word, unless there is at the same time a scientific evaluation of the consequences of the choice and a justification of the choice itself by rational reflection. That is why science policy is not only a scientific practice, but also an art of selection. It is of the essence of science policy to claim for itself to be an art of rational selection, that is, at the same time both enlightened and responsible.

The scientific approach allows one to make enlightened choices. It cannot however, of itself, make the choice completely responsible. For the man who makes the choice is only truly responsible, only really capable of taking upon himself the responsibility and answering for himself in the face of his own judgement and that of others (including those as yet unborn), to the extent that he really makes an effort to put himself at the level of the deepest demands of the action.

But the action is not obscure to itself: it is capable of being aware of itself as it comes to grips with what is happening, in its material involvement which is always specific, and in this way to reflect the demand which urges it on. Indeed this demand is nothing other than the voice of reason in us. The hazard of the action is the hazard of reason.

Perhaps one could make the status of science policy clear by distinguishing between the terms 'rational' and 'reasonable'. The scientific method is rational in so far as it measures knowledge against an intersubjective norm of relevance and efficiency which defines an ideal of operative knowledge. The demands of common sense—or reason—that is of harmony, or of freedom, determine the sphere of what is reasonable; and it is to this sphere that the justification of the norms belongs.

The peculiar mission of science policy is to combine the rational with the reasonable, by putting at the service of the latter all the resources of rationality both in the shape of a scientific justification of the decisions, and at the same time a reflective justification of the norms in accordance with which the decisions are taken.

The responsibility of science

Science policy brings about the penetration of science into the sphere of power. At the same time it increases the responsibility of science. Today we can no longer consider science as a simple external tool in relation to the ends which we have set ourselves. It is no longer possible to distinguish, in its case, what belongs to the means and what belongs to the ends. In the future, whether we like it or not, man's destiny, the evolution of reason within him, comes about through the development of science. Science can no longer think of itself solely in terms of knowledge. It must henceforward think of itself in terms of responsibility. Its evolution is thus bound up with the ethical evolution of man.

Of course, this last is the concern of all men. There is need for participation, discussion, consultation, the co-ordination of perspectives, so that the orientation of the action may have the chance to develop in the most productive way. But those in control of science, because they know better than others what is at stake, and because their action is effective, are charged with a very particular responsibility. Their task demands not only knowledge, but also wisdom; not only a nice comprehension, but also a harmonious vision; not only a passion for what is rational, but also a passion for what is reasonable. Nothing is unchangeable, nothing is truly complete in itself, nothing is absolutely certain. Man creates himself in uncertainty and danger. He creates for himself a whole vista of exhilarating challenges.

It is on man that the issue depends: either evolution will be obstructed and potentialities not yet unfolded will be turned back on to barren objectives such as productivity for its own sake, or power, or entertainment; or else, on the contrary, creation will be sustained—that is, there will be a real development of man's situational status, and a firmer acceptance by man of his own responsibility for human reality.

We shall no longer be what nature, whether within us or outside us, or some inscrutable destiny, chooses to make of us. Our evolution will be what we ourselves have willed. This much at least, the present mutation of society —still incomplete, but already irreversible—allows us to discern.

Part two

The facts

3 The social, cultural and economic data for a science policy

THE WORLD PLAN AND THE NATIONAL PLAN

In view of the increased speed at which learning and techniques are spreading, it has become clear to everyone that the whole of humanity is called upon to experience development based on science with a growing sense of unity. It is therefore possible that one day we shall have a world science policy, whose object will be to select the most important orientations of research for the fulfilment of mankind's aspirations, and to derive therefrom the main policy for the appropriation of the scientific resources of the world. Could we not even now keep up a permanent international debate on the respective importance for mankind of the conquest of space, the curing of cancer or the solutions of the problems of starvation or urbanization? But we are hardly entering the age when such discussions will lead to important political and financial decisions.

The science policy of today operates essentially on the national plane. Its object is to increase and mobilize the scientific and technological potential of a people or a State in the service of the ends which its government pursues. When several governments share the same objective, it sometimes happens that they make a communal use of their resources. International scientific co-operation, which is the subject of Chapter 8, is without any doubt the beginnings of an evolution towards a class of incentives which is wider than the national interest. But it is important to state that, at the present stage, the ends of this co-operation remain strictly national, and that the institutions created for it are not successful, except when the national interests of the participating States find in it their proper reward.

Now it is clear that the total of the national interests, in matters of science as in everything else, does not coincide with the interests of humanity and, therefore, that adding up all the national science budgets would not give us any idea of a reasonable science budget for the world. Everyone knows that, out of this total, research for military ends receives a very large part, and that projects whose incentive is industrial competition and the struggles for prestige absorb the largest share of the rest. Very meagre, in comparison, is the share allotted to medical research, to the preservation of our natural and cultural patrimony, to the problems of the economic development of the Third World, to difficulties such as those arising in connexion with urbanization, education, the pollution of air and water, and so many other urgent problems of the age in which we live.

The intellectual regrets this distorted thinking, while considering it inevitable for governments and industrial enterprises to be less interested in the kind of research whose results will benefit the whole world, than in research from which they hope to derive power, prestige or profit in the competition which faces them. It is therefore left to the United Nations, and to the different institutions connected with it, such as Unesco, to make good, to the limit of their resources, the most urgent deficiencies among those which are left. That is why they are directing their essential effort towards the concrete contribution which science can make to the economic, social and cultural development of all the nations. It is useless to emphasize how very disproportionate the funds actually available are to the needs of this kind. With the purpose of using these in as rational a way as possible, the whole of the efforts of the institutions which belong to the organization of the United Nations will, in the future, be concentrated on a 'World Plan for the Application of Science and Technology to Development' in which perhaps will be seen the germ of a future science policy at the world level. But for a long time, no doubt, this will include only a small fraction of the resources which the world devotes to science.

Since today the accent remains on national policies, it is right that the Science Policy Division of Unesco should make a determined effort to help Member States to work out or complete their appropriate policy in the field of science and technology. Over a relatively short space of time, Unesco has acquired considerable expertise, derived from the comparison of the experiences of countries in every stage of economic, social and cultural development.

This has brought to light the fact that the objectives, methods and plans which it is suitable to recommend to a country for its scientific and technological policy, depend very closely on the phase of development which it has reached, and that they cannot be reduced to a simple or universal typology. This essay will have fulfilled its purpose if it provides guidelines enabling the authorities of any country to analyse its situation, and from these premises to plan the framework and the system of procedure relevant to its needs.

THE ECONOMIC CONTEXT

The situation of the countries of the world is infinitely diverse; not one of them, in its economic development, follows exactly the same stages as those which have preceded it. However, every country passes through a series of successive stages which form a logical and necessary chain and which show kinships or analogies with those through which other countries have passed. It is therefore useful to refer to some typical situations in order to analyse and appraise the framework and situation of an economy, and to initiate a study of the objectives which the country could pursue and the policy it could adopt in consequence.

For the description of these situations one could choose a kind of 'evolutionary' order, proceeding from the most simple stage to the most complex stage of development. The reverse process is less Cartesian, but it makes it more possible to grasp the essential differences from the beginning. This is the way that we have chosen.

THE UNITED STATES OF AMERICA

The most advanced stage of the application of science to development is found in the United States of America. This country has achieved a *per capita* gross national product (GNP) of about 3,700 dollars.

Agriculture, thanks to thorough mechanization, does not employ here more than 6 per cent of the working population. In the manufacturing industries, mechanization and automation are also setting free a part of the labour force, which finds its way into the services sector, in proportion as the essential needs of the population for manufactured goods reaches saturation point.

Inside the manufacturing industry itself, a part of the population is progressively shifting to mechanical and electrical construction, since the most important sphere of production of the industrial worker is less and less the manufacture of a product, or even the servicing of the machine which manufactures it, but more often the construction of a machine which will in future manufacture the product automatically.

Corresponding to this profound change in the structure of industrial labour, there is an equally radical transformation in the nature of the work and the qualifications which it demands. This evolution, for reasons which will be summarized below, corresponds with a need for a fuller general education for a larger proportion of workers.

Let us begin with agriculture, although this sector has ceased to play a characteristic role. As the number of active workers per farm is constantly diminishing, the worker must continue to be able to fulfil a whole series of diverse functions, despite their increasing technicality. The number and level of the skills necessary for an agricultural worker are therefore progressively rising.

In contrast to the agricultural worker, the industrial worker only carries out one kind of task, but the proportion of simple tasks to complicated tasks is changing rapidly. The simplest manual tasks are disappearing the first, and this will be soon followed by the mechanization of difficult manual work, which once used to be the pride of the skilled artisan. The routine intellectual jobs, which employed countless office workers, are progressively being handed over to the computers. The technical activities of repair and maintenance of the plant, which do not diminish, are becoming predominant in industry. The work of management, sales and planning for the future are also no longer the same. The plant becomes more complex, the product and its marketing change more quickly. Work is no longer strictly repetitive but demands initiative and judgement.

The service sector is undergoing an evolution parallel to that of manufacturing industry, and also resembles it more and more in the use of machines, the increase of the wage-earning staff and the decrease in independent tradesmen.

Whatever his work may be, a workman can no longer count on carrying out the same kind of tasks for the whole period of his working life, because jobs have lost their permanence. Between the beginning of the working life of a workman or employee and his age of retirement, successive technological advances will have upset many times the methods of production. Workers are compelled, therefore, to adapt themselves to new tasks many times, even if they remain in the same industry or in the same sector. There are many who have to change their sector in the course of their career.

In the science-based industries of which the sector of capital goods is a part, the product is subject to even more rapid changes than in the other industries. That is why an increasingly important part of the labour force of these sectors is not employed in current production, but in preparation for future production: scientific research, experimental development, technological innovation and the commercial marketing of new products and processes: the sum total of these operations occupies an increasingly prominent position in the activities of firms whose product has a commercial life lasting only a few years. This is especially the case in the chemical production of synthetic materials, and also in the case of machines and machinery; new products and new models which are the results of research, are destined at the end of a few years to replace those which are produced and sold today. The proportion of personnel in these industries whose work demands a high standard of general intellectual training tends to increase rapidly.

To sum up, the decline of simple physical labour (manual workers) has been rapidly followed by a decline in the use of purely manual skills which were acquired by apprenticeship (specialized manual workers and skilled artisans) and of the ordinary office jobs (clerks). Specialization is still as necessary as ever, but it depends more on a stock-in-trade of theoretical knowledge—without which the workman cannot either easily adapt himself to this task, or 'recycle' himself periodically, or—even less—readjust himself when he has to change his job. On the other hand, the process of research and experimental development, technological innovation, and the propagation of the innovation which naturally involve a minimum of routine tasks, can only be entrusted to a person of intelligence. It is the same in the business of management, which gains in importance as society becomes more complex and diversified.

The brief summary above explains why the present economic structure of the United States demands the education and training of a considerable proportion of the working population to a high level. The spread of secondary education and the development of higher education have therefore corresponded in this country to a direct need of industry.

The demand for education is made still greater by the fact that the United States has been led to increase specialization on products of advanced technology, and thus to become the provider of capital goods and synthetic products for a large part of the world's needs. This specialization is the result of the conditions of competition prevalent on the world market, between countries whose *per capita* GNP is very different.

It is known that the average pay of workmen is, as nearly as possible, proportionate to the average *per capita* GNP reached in the countries under consideration. In the traditional consumer industries, like the textile industry, or in the basic industries, like iron and steel, the countries which pay high wages are finding that competition from the newly industrialized countries is becoming very keen. American industry tends, therefore, to slacken production in these industries in order to specialize in those where its progress in technology and education assures it a protection which, while naturally temporary, is on the whole fairly secure. These are the science-based industries.[1] The developing countries will only reach high rates of educational enrolment at the secondary or higher level after several decades, and it is only then that they will have at their command the necessary numbers of educated workers to start in their turn a quantitative expansion of science-based industries. Of course they already occupy some positions in these industries.

Thus it can be seen that the place occupied by science-based industries in the economic organization of the United States is larger than its internal needs alone would justify, and that the rigours of competition are driving industry and the State to lay even more emphasis on this specialization, by deploying scientific research activities and experimental development in all the orientations of advanced technology.

It can therefore be stated that the stage of economic development reached by this country has produced there a 'structural' need for a large scientific potential.

But it has come about that the international situation caused by the 'cold war', and the race between the Soviet Union and the United States for the possession of an absolute power of annihilation, has provided the Government of the United States with incentives for an exceptional financial effort in favour of science. The outer-space competition has taken over since 1957. Although it is not military in its purpose, space travel demands technological achievements in propulsion and automation, and the strategists know that in these fields it might be dangerous for their country to fall behind. The incentive of prestige and considerations of caution in face of the technological progress of the rival power could have motivated policy adopted in this field.

1. In aggregate these can be taken to mean the sector of mechanical, electrical and electronic construction, and the chemical industry.
   In some of their forms, mechanical or electromechanical or electronic articles have already become objects of keen competition, especially from countries—such as some European States, Japan, etc.—which have reached an intermediate stage of development. But this is chiefly in mass-produced articles intended for household use, whose technology has become standardized. It is typical that, even in such sectors, the United States has been compelled, by the competition of countries paying lower wages, always to take refuge in more advanced techniques.

It is therefore political rather than economic reasons which have brought about unprecedented financial support for industrial research by the State. Such financing always coincides with the 'structural' need mentioned above, a need of which J. K. Galbraith has given a masterly analysis in *The New Industrial State*. Political motivations came in at an appropriate moment to provide a fresh impetus to the American economy.

The huge State programmes directed towards nuclear energy, computers, aeronautics and space have thus been able to serve as pathfinders to all science-based industries. They have forced the majority of the other industrial sectors to achieve, in their turn, scientific and technological breakthroughs by the new problems which they set them. The case of scientific instruments is characteristic of the effects of this drive. On the opposite side, the case of the automobile industry is typical of the slowness of technological evolution in the absence of this incentive.

In this context one must make a special mention of electronics and its newest and most spectacular application: computers. Under the impetus of huge State programmes and under pressure of the needs of nuclear, military and space research (the first requiring powerful mathematical processing, the second demanding miniaturization and high reliability in logic circuitry, and the third involving construction of large ultra-fast computers) electronics has accomplished two technological mutations in ten years: the transistor and the integrated circuit. The side effects of these breakthroughs, not only on the whole of the advanced industries but also on various sectors of the services, have already proved conclusive; but their most important consequences still continue to be felt as the use of computers improves and becomes more general.

It is therefore highly probable that the powerful resources placed at the service of scientific progress in the last ten years have induced the United States to make a new start after a period of semi-stagnation, and to occupy once more very strong positions in foreign markets. By way of direct investment, American businesses have even been able to obtain strong positions inside the economies of several countries.

Taking the United States as our example, we shall now try to define the characteristics of the scientific potential of a country during its second phase of industrialization.

These characteristics would be the following:

1. Technological research in the science-based industries occupies the first place. Fundamental research in physics and chemistry, which is essential for the process of innovation in these industries, would equally be the object of a large deployment of scientific activities and resources.
2. The State takes responsibility for a considerable proportion of the cost of these researches, both fundamental and technological, when the occasion arises for big national programmes; this results in a close co-operation between the industrial enterprises and the government, the last instead of the first taking the greatest part of the risks of technological innovation, by an abundant and constant flow of contracts for research and experimental development including orders for advanced technological equipment.

3. Higher and technical education spread rapidly, because the economy needs manpower with practical knowledge, who have received an extended general education, and because of the demands of the laboratories for research workers. The university therefore holds a vital place in the national life. To its role in education and training, it adds that of executive centre for an important part of the oriented fundamental research and of the applied research. Generous contractual aid is thus given to universities by the agencies which administer the large governmental programmes.
4. The relative importance of agriculture and the traditional basic industries in the economy continues to diminish; their chances of making a profit decrease gradually, inducing them to claim protection and subsidies.
5. The habit of including large budgets for research in their development programmes spreads from the sectors of defence and science-based industries to the other sectors of national life: there it triggers off 'systems analysis' studies and 'technological forecasting' exercises which are in fact attempts to generalize the application of science to development. Management methods have been adjusted in order to 'organize the change' or to 'plan the technological innovation'. The uneasiness caused in many people's minds by the fear of an unsettled and unknown future disappears when this future takes a definite shape, and when it becomes clear that planning will be able to control or counterbalance its consequences. People's attitude towards the future then becomes more receptive.

To sum up, the science-based industries, and consequently science itself and the institutions which pave the way for these industries, have taken a central place in the economy of society. Research and experimental development absorb more than 3 per cent of the GNP. There are three research workers for every thousand inhabitants; one-third of young people from 20 to 24 years old are pursuing studies. The scientific attitude of mind tends to spread to all activities. Systems analysis and planning for change are the most characteristic cultural signs of its impact.

Some authors have extrapolated this evolution and imagined a society where industry (even in its most scientific forms) would have ceased to occupy the front of the stage, and where the centre of gravity of scientific expenditure would have shifted towards the application of the scientific process to the objectives which we defined in the first chapter as raising the quality of life and the quality of society: control of one's environment, urban technology, applied biology, applied sociology, and tertiary economic activities. This type of society has been called 'post-industrial civilization', and it is clearly a vision for the future: the American reality today, as much internally as on the world stage, still remains largely dominated by the needs and interests of industry.

OTHER INDUSTRIALIZED COUNTRIES

The description which follows gives a sketch of the situation in the countries of Western Europe and the U.S.S.R. These conutries have many characteristics in common with the United States, although certain structural characteristics of the new phase, into which they have entered, have still not reached their full stature. In many respects Japan ought to be considered as a country in this group because of the speed of its growth and evolution.

Here we are dealing with countries which have completed their first industrial revolution and whose population is mostly urbanized and has already been completely educated at the primary level. They are at the moment carrying out the second phase of their industrialization.

There, the science-based industries are in full expansion and are employing a growing proportion of the working population, while agriculture—as an employer—continues its retreat at a rapid pace; service industries are correspondingly developing. Although the mass consumption of cars and electrical appliances is becoming general, thanks to the rapid progress of industrial automation, the centre of gravity of the economy often remains in the traditional basic and consumer industries.

The production per head is rapidly growing, while remaining, on the average, appreciably inferior to that of the United States (which is of the order of 1,000 to 2,000 dollars per head). A smaller productivity in these countries is usually caused by the survival of unsuitable industrial structures, of persisting low investment of fixed capital per worker, and of outdated methods of business management which remains inhibited or bound by tradition, especially in matters concerning technological innovation. The techniques used are not distinctly less advanced than those which one meets in the United States, but the dissemination of the knowledge and the propagation of recent technological advances suffers delay in these countries. The number of computers is a striking indication of such delay.

In world trade these countries are, like the United States, exporters of capital goods. Some of them, however, have taken the place of the United States on the world market as providers of industrial semi-finished products and textiles, while expecting to be replaced, in their turn, by the next wave of countries which complete their first phase of industrialization. The competition of the latter is very keen, and is already forcing European countries to efforts of rationalization, which lead to a heavy loss of jobs in the traditional industries. The resulting social situation compels these countries to speed up the conversion towards mechanized and chemical industries, which the growth of their internal needs requires in any event.

On the world market of machinery, vehicles and appliances, the European and Japanese electronic industries maintain their competition successfully in goods that have already become traditional, like automobiles, ships, domestic appliances, etc., thanks to lower manufacturing costs than those of the United States. When the size and organization of the enterprises are adequate, it even happens that they completely outclass American competition, especially in different fields of 'little technology' where the United States dominated only twenty years ago. But these countries present a less striking figure in comparison with the United States in matters concerning the most advanced products of 'big technology': nuclear power-stations and fuel, aeroplanes, computers, etc., and in matters concerning all the fittings and components whose improvement is the direct result of national programmes of research, experimental development and technological innovation (scientific instruments, electronic integrated circuits, rare or special metals, etc.).

The structure of the scientific potential of Europe and Japan already shows many of the same characteristics as that of America, but in a lesser degree:

1. The development of new products in the chemical and electromechanical industries already absorbs more than half of the resources. However, agriculture and the traditional industries still have considerable importance, and this shows itself in the size of the research institutes which specialize in these products, and in government centres or co-operative research stations, whose activities absorb a still considerable share of scientific resources. They alleviate the relative withdrawal of these industries, by slowing down the decrease in their profit-earning capacity.

2. Large national programmes of research and experimental development play an important role in certain countries: the U.S.S.R., France, the United Kingdom and, more recently, the Federal Republic of Germany and Japan. But in smaller industrialized countries such national programmes are of little importance. However, the large countries themselves (with the exception of the U.S.S.R.) are less and less able to support a policy of 'big technology' and to exploit its results commercially. Industries based on advanced technologies suffer from the more rapid progress of their North American competitors who enjoy greater government support and also take better advantage of scientific methods of decision-making. Their second phase of industrialization is slowed down, and problems of employment arise. American mechanized, electrotechnical and chemical enterprises therefore establish numerous important subsidiaries in these countries, when governments permit (this is the state of affairs in Western Europe; it happens less often in Japan and never in the U.S.S.R.). But this kind of industrialization is not necessarily accompanied by a strong growth of research activity, for the main research effort usually continues to be directed from the parent organization, on the other side of the ocean. The more restricted the resources of a national State are, the worse the effect this phenomenon produces.

3. Higher education and higher technical education have undergone a sudden expansion in the fifties; this naturally entailed crises of growth and adaptation. The process of mass education and training at the higher level has begun.

The situation described above is that of a country whose second phase of industrialization is well under way. Countries which are hardly approaching that stage, or are only now reaching the end of their *first* phase of industrialization, should also be mentioned. Here we are dealing generally with countries of small or middle size.[1] Their scientific potential appears at first sight —if one only considers the percentages—to be directed more towards the sciences with a humanitarian purpose, like medicine or sociology. But this is a distorted view which springs from the small volume of research activities

---

1. Some of the large developing countries (Brazil, India, Indonesia, Pakistan) have obviously not reached this stage either.

in their national industry, as well as from the absence of large national research and experimental development programmes. In fact non-oriented research does not absorb a larger proportion of national resources in these countries than it does elsewhere.

COUNTRIES IN THE FIRST PHASE OF
INDUSTRIALIZATION

A large number of countries in the world are at the present time in the first phase of industrialization.

This phase of their development has begun, for many of them, after a period of contact with world markets while under colonial status. This caused their commercial activities to be exclusively geared to foreign trade: the export of raw materials and the import of all the industrial products necessary for the country (consumer goods, intermediate goods, and capital goods).

The *first phase of industrialization* aims chiefly at replacing, by means of local production, the import of consumer goods and intermediate goods. In the course of this phase, the country remains almost entirely dependent on imports for the capital goods necessary for the build-up of its industry, and also for highly specialized chemical products. It pays for these imports with currency earned by the export of renewable or non-renewable natural resources, but also—and in growing proportion—by the export of intermediate goods (steel, non-ferrous metals, cement, glass, common chemical products, etc.) produced by the young national industry. It also becomes an importer of certain raw materials which it lacks but which are needed for its industry.

During the first phase of industrialization, industrial technology is for the most part imported from abroad: either the country accepts foreign firms, or it buys patents or licences, or again it enters into contracts with builders of 'turn-key' factories—contracts which cover starting-up expenses and the technical training of the necessary local personnel. The national effort towards original technological innovation, and the supporting programmes of research and experimental development, therefore aim primarily at adapting foreign techniques to the characteristics of the indigenous raw materials or to the tastes and habits of the local customer. When the products are destined for export, research effort directed towards the *improvement of the quality* becomes a pressing need, because of the competition that prevails on the world market.

Moreover, it is constantly observed that the *horizontal transfer of technology* only succeeds properly with the intervention of the research teams of the recipient country. Without having to deploy a massive creative effort, the countries in the first phase of industrialization are nevertheless led to maintain a certain research potential in all branches of technology which they have decided to introduce into their own economies. Co-operative research centres whose membership includes all enterprises of a given industrial sector, or institutes of technology, often play this role.

With this reservation, the main effort of applied research of countries heading towards industrialization generally relates to agriculture. Experimental agronomy, which is almost entirely dependent on local conditions of climate, soil and biological environment, is liable to suffer harm by the horizontal transfer of technology, since the majority of the scientifically advanced countries belong to temperate zones, while the majority of the other countries are from the arid or humid tropical zones.

Apart from the agricultural research stations and some centres for technological research, the scientific potential of countries in the first phase of industrialization tends to be concentrated in universities, schools of medicine, and institutes of higher education in technology and agriculture. The scholars there enjoy the high prestige which surrounds intellectual activities in traditional societies. This prestige is frequently associated with cultural refinement, and with a rather rigid stratification of society.

The universities of countries where the electromechanical and chemical industries have already taken a central position in scientific research and in the economy, take a considerable interest in applied science, because of their constant contact with these industries. On the other hand, universities which lack such contacts are attracted by the theoretical aspects of pure research, since this seldom requires costly instruments. The universities of the developing countries often reach high standards of excellence, and some of their researchers are successful in winning honour for their people, thus giving a nation confidence in its creative power. However, such an occasional honour hardly contributes to the national development. In fact, fundamental discoveries made in this way cannot be exploited by the country which makes them. It possesses neither the units of applied research, nor those of experimental development, nor entrepreneurs capable of industrializing and marketing a decisive invention. In practice, the scientist who made the discovery will be honoured by the fact that some important American review publishes his results without delay. These results will then be scanned by several industrial laboratories, which are looking out to find in basic sciences some clues leading to fruitful technological innovation. With the managerial staff of their company, the directors of these laboratories will mount plans for an industrial and commercial operation. In the end the country which originated the discovery may well have to buy the licence for its application. Alternatively, it may have to admit within its borders a subsidiary company of the foreign industrial enterprise which has developed the original discovery.

While the contribution of the university to the progress of national industry is thus limited, its contribution to intellectual and cultural progress is always less effective than it ought to be. The intermediate grades between a scientific élite of the highest standard and the still half-illiterate masses are often insufficiently filled, although it is through these medium-level specialists and technicians that modern knowledge and attitudes spread through the social structure of a country. In certain countries, prejudices of a political, social or religious order even tend to aggravate such academic isolation.

However, several countries have succeeded in making the academic world participate more fully in the general progress of the nation by means of a policy which directs the energies of the university, not only towards the highest achievement in pure science, but also towards the large-scale education of qualified cadres at advanced level, and even at medium level, for industry and the economy. Since the first phase of industrialization demands considerable numbers of engineers, technicians, economists, etc., these countries sometimes prefer to entrust their education to the university rather than to higher technical schools. Standards of teaching in the applied sciences have thus been raised, and closer liaison has been secured between the university and the most modernized section of the nation.

It must also be noted that, during the first phase of industrialization, the *scientific public services* are called upon to play an important role, since they form the infrastructure necessary to all modern activities—industrial and others. The mapping of the national territory and the survey of its natural resources by means of the most modern methods in geology, geophysics, hydrology, seismology, climatology, pedology, human geography, ethnology etc., are clearly of decisive importance. In fact these sciences provide the indispensable data for the setting up and carrying out of the national plan for industrial and agricultural development. The national library and the national network for scientific and technical information constitute on the other hand essential elements of the *infrastructure of auxiliary scientific institutions* needed for the successful horizontal transfer of technology and for the penetration of science into every sphere of the national life. Finally metrology, scientific and technical standardization, laboratories for analysis, experiment and testing, constitute a third sector of the infrastructure of scientific institutions necessary for an industrial country during its phase of accelerated growth. Yet this infrastructure is far from being always complete.

THE STAGE WHICH PRECEDES INDUSTRIALIZATION

This economic stage, as we have already pointed out, is characterized by the export of raw materials obtained by mining, agriculture or forestry, and by the importation of almost all the industrial products which the mining and oil companies, the plantations, the State and the wealthy landowners need. The basis on which the landowners' power rests often proves an obstacle to the implantation and growth of industry and to the vigorous agriculture which characterizes the nations already on the road towards industrialization (see page 68).

The need of managerial staff in the national economy is therefore limited by the underdevelopment of the manufacturing industry and the tertiary sector. As a consequence, higher education is poorly developed, and mainly focused on literature, law and medicine, though it sometimes happens that there are also some active nuclei of pure science. In short, academic structures are like those in Europe up to the eighteenth century. It is only by courtesy that this group of countries is said to be 'developing'; in fact, the term was coined to describe the position of the countries in the first phase of industrialization.

THE CONSTRAINTS WHICH SPRING FROM THE
ECONOMIC AND SOCIAL CONTEXT

It is clearly impossible for a nation, in ten years, to leap over the gap which separates the fourth group from the first or even the second group of countries described above: not only must the industrial structures and plants be created—and the cost of these bears no relation to the present GNP of such a country—but besides, almost the whole population must be converted to activities for which it has had no intellectual training. The need is therefore felt for a sudden cultural adaptation to situations and behaviour proper to an industrial civilization. For that purpose it would be necessary that about a half of the active population be given educational training at the secondary level and one-tenth at the higher level. The resulting delay proves to be an infinitely bigger constraint than the problems posed by the accumulation of fixed capital, since capital goods can be produced abroad or made available by means of programmes of technical assistance, loans or direct investment. Unfortunately none of the existing methods of adult education will make it possible to accomplish the whole of this process in less than a complete generation. In fact, the countries whose development is most rapid at the present time will certainly devote more than twenty-five years to it.

It is reasonable to forecast that the GNP *per capita* of the developing countries will rise exponentially during the next decades. The pattern of employment and level of education of the active population will change in order to adapt to the needs of the most important or characteristic productive activities, while the level of consumption of the people will rise. In the course of this process, the scientific potential of the nation ought to increase considerably, and the operational network of scientific and technical institutions ought to change its dimension, structure and missions. Such appear to be the constraints which the governments of the developing countries inevitably have to face, if they decide to increase to the maximum the pace of their economic and social progress.

There are, however, other obstacles which with the help of the appropriate policies these countries will be able to surmount more easily, because these obstacles do not reach to the very foundations of the process of development but concern rather the superstructures peculiar to different phases of industrialization. These are the psychological, social or political resistances to certain kinds of change, and especially the perpetuation of mental habits and behaviour or the survival of institutions that belong to a bygone age. Such inertia often checks the adaptation of attitudes of mind and behaviour to new situations. In extreme cases it can bring the process of development to a halt or even into regression.

Fortunately, the universality of the process of development, which now seems to be self-sustained thanks to the momentum it has gained, should leave no anxiety as to the possibilities of such disastrous events. It is possible that for a given nation, considered in isolation, the check in the development may last several decades, before the sociological or political barriers give way under the pressure of aspirations temporarily repressed, and of the progress made abroad. Meanwhile, other nations, more adapted to change, will have taken decisive steps forward.

The uneasiness which can spring from the perpetuation of unsuitable superstructures and behaviour is therefore understandable, especially when it manifests itself in that part of the social system which ought to be the most 'militant': the scientific and intellectual élite of the nation, and especially the academic community, high administrative circles or among industrial entrepreneurs. Some people are tempted to rank this perpetuation as a constraint, so disastrous can its influence be on the progress of the nation. It is however no more than an obstacle, since appropriate political action is capable of sweeping it away.

When considering the survivals or perpetuations opposed to general economic development, one must obviously rank first the resistance to organizational change in rural communities: this is the main and most common cause of economic stagnation in the developing countries. Naturally, the absence of sufficient or adequate industrial entrepreneurship can also cause a check in both the first and second phase of industrialization. Such barriers may result either from the weakness of the national structures or from foreign domination, or from both at once.

At the level of the administration of the State and of big organizations, the most typical anachronism is undoubtedly the refusal to abandon traditional management-through-regulations-and-supervision in favour of integrated management-by-objectives which leaves to the executive grades the task of optimizing the performance of the units of production put at their disposal.

Freedom of action can obviously only be granted to those executive grades if a framework of clear objectives has been defined, and it must be accompanied by an effective responsibility of the executives as to the results obtained. Higher management must therefore be able at all times to know, to follow closely and to assess the management of the lower grades, without however allowing itself to intervene except to make clear its objectives and evaluate the results. Such a rule imposes, even on the government, accuracy in the formulation of objectives and in the assessment of the results—an accuracy which administrations of the traditional kind did not need, and for which, indeed, they possess neither the methods, nor the machinery, nor even the intellectual background and necessary information. Accustomed to act in a simple and almost static operational environment, on the basis of standardized instructions derived from normative typologies, traditional administrations developed a rigid system of planning, which took no account of actual situations or of their shifting complexity. That system remained virtually unchanged from the age of Augustus to our own, and shows itself unsuited to the diversity of situations, products and techniques which sprang from the economic development of the twentieth century.

Today, the rapid transmission of analytical data, and their processing by computer, enables the summit of the hierarchy to know and assess the complex situations which develop at the periphery and therefore to allow executive cadres the freedom of movement which they need. In modern management, the independence of the executives entails scarcely any risk to the cohesion of the whole, since the leadership has at its command techniques of information which allow it to follow and analyse, in detail and day by day, the progress of every managerial entity, whatever the degree of decentralization may be, and to intervene before things deteriorate. The centralization of administration has therefore lost its essential purpose, which was to ensure a monolithic solidity of authority, to which efficiency and economic optimization have often been sacrificed.

But the prerequisite choice of objectives demands a quantitative and qualitative forecast of future circumstances as well as a knowledge of the variables of the situation, and of the parameters of constraint and coherence of the system. Macro-economics, econometrics, programming, operational research, and systems analysis are the necessary tools of such a 'futurology'. Management supervision has also recourse to mathematical and statistical analysis of the actual situation.

But none of the hierarchies accustomed to a standardized management of the legalistic and disciplinary kind possesses these techniques. Moreover their first reflex is usually an authoritarian reaction, aiming at silencing at the same time those in professional jobs or in the intelligentsia, who propose that the nation should give itself more valid and less simplistic objectives, and those in executive office at operational levels—heads of businesses or services, schools or laboratories—who demand more independence so that they can contribute to achieving such objectives by a more efficient use of the resources at their disposal, resources which they know are being wasted every day.

In contrast to the authoritarian survivals of the hierarchies can be placed the survival of individualistic attitudes and the resistance of certain groups to any kind of organization. They are composed of groups or categories which have known some professional success in the 'atomized' socio-economic structures of the past; the so-called liberal professions, higher education organized around the 'professorial chair', individual theoretical research which did not require any important logistic support. The survival of attitudes of individual competition, exacerbated by some traditional notion of the honourable status of 'the boss', is obviously an obstacle to the setting up of multi-disciplinary research teams working towards the same objective. The survival of opposition to organization and planning is equally an obstacle when some complex operations are undertaken which, like the monitoring of observations from automatic space satellites, entail the simultaneous solving of various problems without allowing the failure or slowness of one of the researchers to compromise the success of the whole operation.

The contempt of old-fashioned scholars for the financial and economic aspects of their activities is no longer compatible with the vast sums and equally vast hopes which the nations today invest in research. The same thing holds good, for the same reasons, of the organization and financing of higher education, the efficiency of which conditions the future of the nation in a way that was not realized in the past, when the university was above all a privileged place where flights of the mind found a refuge from the contemporary scene.

The importance of the stake is today perceived by the young generation of intellectuals with the same pitiless gaze which pierces the anachronism of structures and attitudes. No doubt that is why protest is unleashed today against bureaucratic or academic obstruction. It seems that one must count on the pressure exerted by youth, as much as on the maintained and persistent action of the government, to remove these obstacles one by one.

Foreign domination is another impediment on the road to development. We are not speaking here of political or military domination which countries or groups of countries can impose on others. We are thinking rather of certain factors of an economic or technological nature, springing from a foreign volition, which can weigh heavily on the development of certain nations. This is the case when the raw materials which they produce are bought at too low a price, or when obstacles are placed in the way of their processing on the spot. Such a situation clearly deprives the producer country of the capacity for investment which its growth requires. In these conditions the development of the scientific potential of these countries—like that of their higher education—can become, paradoxically, a source of disharmony and impoverishment, especially when its effects lead to the emigration of élite personnel.

4 The resources devoted to scientific research and experimental development in the world

The information available on the whole of the resources devoted throughout the world to scientific activities is still fragmentary and inaccurate. With few exceptions, only the industrialized countries have reached the stage of making a systematic survey of their scientific potential.

The statistics that are presented in this chapter ought therefore to be handled with certain reservations, in particular for countries whose scientific potential cannot be easily assessed. Estimates have had to be made to complete the available data. The risks of error which these estimates entail are considerable indeed when they deal with countries as vast and densely populated as mainland China.

Nevertheless these estimates make it possible to give an idea of the general distribution of scientific resources in the world and to appreciate, with regard to size, the considerable differences which exist in this respect between the different regions.

It is true that the record of the resources devoted to science is only rather a rough gauge of the level of the scientific activities of a country or group of countries. It ought to be completed by a study of the productivity and effectiveness of the various national scientific potentials, and for that purpose to call upon elements of a qualitative character as well as quantitative ones.

In the present state of the international statistics of science one has to be content with purely quantitative data. These seem however to be sufficient for the object of this chapter, which is to show the inequality of opportunities offered to different countries as regards their participation in the vast creative process that springs from the rapid progress of science and technology.

In fact, numerous studies tend to prove that an over-all connexion exists between the level and concentration of resources which a country devotes to research and experimental development on the one hand, and its tendency to innovate and create on the other.

WORLD DISTRIBUTION OF SCIENTIFIC RESOURCES

Table 1 shows that 86 per cent of the total of scientists and engineers working in the world are concentrated in a zone which embraces North America, Western Europe, Eastern Europe (including the U.S.S.R.) and Japan. This zone is inhabited by 30 per cent of the population in the world. If one reckons only that part of this category which is devoted to research and experimental development (R & D), the concentration appears even more powerful (94 per cent).

TABLE 1. The number of scientists and engineers in the world (about the middle of the decade 1960-70)

Regions	Total number of scientists and engineers in the world		Scientists and engineers engaged in R & D		Population	
	(thousands)	(%)	(thousands)	(%)	(millions)	(%)
United States of America	3 500	24	497	28	195 500	6
U.S.S.R.	4 800	34	637	35	258 750	8
Western Europe	2 000	14	264	15	310 000	9
Eastern Europe[1]	1 400	10	173	10	120 000	3.5
Japan	600	4.2	115	6	98 250	3
Continental China[2]	540	3.8	53	3	690 500	21
Latin America[2]	190-200	1.5	7-8	0.5	166 000	5
Asia[2,3]	1 000-1 200	8	40-50	2.3	1 150 000	35
Africa[2]	60-70	0.5	2.5-3.5	0.2	311 000	9.5
TOTAL	± 14 000[2]	100	± 1 800[2]	100	3 300 000	100

1. Without the U.S.S.R.
2. Estimate.
3. Without mainland China and Japan.

Sources: Unesco and OECD.

TABLE 2. The number of scientists and engineers per 1,000 inhabitants (1965)

United States of America	18.0	Burma	0.1
U.S.S.R.	18.0	Cambodia	0.04
Countries of the European		Ceylon	0.5
Economic Community (EEC)	10.0	India	1.5[1]
United Kingdom	10.4	Iran	0.05
Sweden	7.0	Laos	0.06
Canada	13.0	Pakistan	0.2
Japan	8.0	Philippines	2.7
		Thailand	0.8
Chile	1.8		
Columbia	1.5	Ghana	0.6
Venezuela	1.6	Kenya	0.3
		Nigeria	0.1
Afghanistan	0.3	Uganda	0.1

1. Including technicians.

Sources: Unesco and OECD.

Table 2 shows, for a certain number of countries or groups of countries, the totals of working scientists and engineers as percentages of whole population.

This percentage is the highest in that same group, comprising the countries of North America and Europe and also Japan. In the U.S.S.R. and the United States of America, those with degrees in science and the engineers represent about 20 per cent of the total population. In Europe and Japan the figure is about 10 per cent. In most developing countries the proportion varies from 0.04 to 2 per cent of the population.

Table 3 shows the intensity of the financial effort devoted by certain countries to research and experimental development (R & D) expressed

TABLE 3. National expenditure on research and experimental development (R & D) as a percentage of the gross national product (GNP) and in dollars *per capita*, and the gross national product in dollars *per capita* (figures of 1965 or the nearest year available)[1]

Country	National expenditure on R & D		GNP *per capita*[2] (dollars)
	As percentage of GNP	Dollars *per capita*	
United States of America	3.4	110	3 560
Canada	1.1	22	2 460
Europe (north-west)			
Large countries[3]	1.7	30	1 876
Small countries[3]	1.2	19	1 789
Europe (Mediterranean)			
Italy	0.6	6	1 100
Other countries[3]	0.2	0.8	457.5
Latin America			
Argentine	0.4	2.8	783
Mexico	0.1	0.5	443
Venezuela	0.1	1.2	916
Asia			
Cambodia	0.01	—	120
Ceylon	0.3	—	137
India	0.3	0.4	92
Iran	—	0.9-1.1	240
Iraq	—	0.1	245
Japan	1.4	9	850
Pakistan	0.3	0.2	95
Philippines	0.2	0.3	237
Africa			
Algeria	—	0.3-0.5	210
Tunisia	0.03	0.6	188

1. Countries with a completely planned economy have not been included in this table because of the difficulty of establishing in their case a comparison between research and GNP which would have any affinity with the comparisons fixed for countries with a capitalist economy. The principles and methods of national accountancy are effectively different in the member countries of the OECD and the Council for Mutual Economic Assistance (CMEA-Comecon).
2. For the countries of Latin America, Asia and Africa: gross domestic product (GDP).
3. Averages.

Sources: Unesco and OECD.

as a fraction of the gross national product (GNP) and the amount of dollars spent on research and experimental development per inhabitant. Clearly, the countries which make the most intensive scientific effort are those which enjoy the highest *per capita* incomes. Again, these are the countries of North America and Europe, and also Japan. As a general rule, the comparison shows that the national expenditure *per capita* on research and experimental development is more than proportional to the income per inhabitant.

CONCENTRATION OF THE RESOURCES EARMARKED FOR RESEARCH AND EXPERIMENTAL DEVELOPMENT IN HIGHLY DEVELOPED COUNTRIES

Table 4 shows that in the industrialized countries of Europe and in the United States of America it is the industrial sector which absorbs the greatest part of the national resources earmarked for research and experimental development (about 60 per cent). In the developing countries of Western Europe, however, it is the State which undertakes the greatest share in carrying out scientific activities. This conclusion can probably be extended to cover all the developing countries which do not adopt a totally planned economy. The predominance of the State results from the fact that the overall figures of industrial research are still very small in this group. They do not necessarily reflect a substitution of the State for private enterprise in the functions of research, experimental development and technological innovation.

TABLE 4. Distribution of national expenditure on research and experimental development according to the area where the activities are carried out, 1963-64 (average percentage per zone)

Country	Industry[1]	State	University	Other sectors	Total
United States of America	67	18	12	3	100
Industrialized European countries, members of OECD	61	16	17	6	100
Developing countries, members of OECD	24	65	9	2	100

1. Public and private industrial enterprises.

Source: International Statistical Year for Research and Development 1963-1964, Paris, OECD, 1967.

Table 5, which refers to the distribution of expenditure on industrial research in the different sectors of activities, for the same group of countries, shows clearly that two sectors—that of electromechanics and chemistry—absorb the vast majority of the sums of money devoted to research. It will be noticed that the centre of gravity of industrial research is shifting from the industries of the first generation (groups 1, 2, 3) towards group 5 (electromechanics) as development proceeds.

78

TABLE 5. Distribution of expenditure on research and experimental development in industry 1963-64 (in percentages)

Sectors[1]	United States of America	Industrialized European countries, members of OECD	Developing countries, members of OECD
1. Extractive industries	—	1.82	5.10
2. Basic industries	2.45	13.51	11.03
3. Traditional consumer industries	2.62	8.32	21.64
4. Chemical industry	13.37	25.56	25.92
5. Mechanical and electrical construction	79.91	46.93	28.34
6. Other manufacturing industries, building, water, gas, electricity	1.65	3.86	7.97
TOTAL	100	100	100

1. For breakdown of the sectors see Table 6.

Source: Recherche et Croissance Économique II, Brussels, CNPS, 1968.

TABLE 6. Coefficients of research according to sectors (expenditure on research and experimental development compared with added value) in the industries of the countries of the EEC and in the United States of America, 1963-64 (in percentages)[1]

Sectors	Federal Republic of Germany	France	Belgium	Netherlands	Italy	United States of America
Basic industries						
Ferrous metals	1.8	1.1	*3.5*	*4.4*	*1.0*	0.9
Non-ferrous metals	1.8	*5.1*	7.5	5.0	*1.1*	1.8
Stone, clay, glass	0.4	0.9	2.1	0.9	0.1	2.0
Metal construction	1.9	—	0.8	5.9	—	1.2
Traditional consumer industries						
Food and beverages	0.2	0.1	0.3	1.5	—	0.6
Tobacco	—	—	0.1	1.1	—	—
Textiles	0.4	0.8	0.4	1.6	—	0.2
Clothing, footwear	0.3	—	0.1	0.3	—	0.1
Wood, cork, furniture	0.1	—	0.2	0.9	—	0.1
Paper	0.1	0.1	0.1	0.9	0.2	0.9
Printing and publishing	0.1	—	0.1	0.1	—	—
Chemical industries						
Oil (extraction and refinery)	*2.0*	1.8	1.5	1.3	0.8	4.5
Rubber	1.7	1.2	14.5	3.6	6.4	3.0
Pharmaceutical products	8.6	6.2		26.3	2.2	7.9
Other chemical industries		3.2		12.7		6.5
Mechanical and electrical construction						
Electrical construction	*4.1*	8.8	5.1	8.3	*1.3*	*14.7*
Mechanical (except electricity)	2.4	0.6	1.9	7.2	0.5	5.2
Instruments	4.5	—	—	—	0.6	*11.3*
Aircraft and missiles		40.3		—		64.6
Motor cars and parts	1.6	2.6	1.0	4.1	3.5	7.3
Naval construction		—		4.5		
Other transport materials		—		—		
Other manufacturing industries	1.6	—	0.1	0.3	0.7	1.4
All manufacturing industries	1.9	2.0	2.2	3.9	0.9	6.2

1. The coefficients in italics are those above the average of the country under consideration.

Source: Recherche et Croissance Économique II, Brussels, CNPS, 1968.

TABLE 7. Distribution by sectors of State financial aid for research and experimental development in industry[1] for the more industrialized countries of the OECD (1963-64)

Sectors	Government financing of R & D (percentages)					Share of public funds directed to each industrial sector (percentages)				
	United States of America	United Kingdom	France	Industrialized European countries of OECD	Japan	United States of America	United Kingdom	France	Industrialized European countries of OECD	Ja
Extractive industries	—[2]	—	—	5.86	0.30	—[2]	—	—	0.36	
Basic industries	5.55	0.91	11.37	4.17	0.68	0.26	0.13	2.79	1.22	1
Traditional consumer industries	6.91	0.63	7.77	3.92	0.26	0.32	0.14	1.47	0.93	12
Chemical industries	15.97	0.37	2.79	0.97	0.07	3.84	0.16	1.99	0.89	28
Mechanical and electrical construction	66.54	50.49	42.11	37.37	0.49	95.58	99.50	93.63	91.99	44
Building, water, gas, electricity	—	0.83	0.45	25.45	0.76	—	0.07	0.12	4.61	2
All industries	56.56	35.00	27.08	24.21	0.36	100.0	100.0	100.0	100.0	100

1. All private or State industrial enterprises and industrial research associations.
2. Not available separately.

Source: International Statistical Year for Research and Development 1963-1964, Paris, OECD, 1967.

Finally, Table 6 gives, for certain industrialized countries, the coefficient of research of the various industrialized sectors (comparisons of the expense of research and experimental development with the added value of the sector). Table 6 makes it possible to get an idea of the intensity of research in the most important branches of industry. Like the preceding table, it shows that the effort of research in a developed economy, and therefore its potential for innovation, are concentrated on certain industrial sectors, those of electromechanics and chemistry.

Table 7 shows the part occupied by government finance in the expenditure on research of the industrial sectors in the advanced countries of Western Europe, the United States of America, and Japan, and also the share allotted to each industrial sector in the total of governmental support to industrial research. It sheds light on the decisive role of governmental support in the sector of mechanical construction. The financial contribution of the government in this sector is particularly important in the United States, because of the volume of research contracts allotted to industries of this sector, within the framework of the large-scale military, nuclear and space programmes. France and the United Kingdom—which have similar, though less important, national mission-oriented programmes—are in an intermediate situation.

THE EVOLUTION IN TIME OF RESOURCES DEVOTED TO RESEARCH AND EXPERIMENTAL DEVELOPMENT

Table 8 describes the recent evolution of expenditure devoted to research and experimental development in a certain number of countries whose data it has been possible to collect. It shows that the growth of this expenditure has been rapid everywhere.

Information available concerning the evolution of the countries which are beginning their industrialization is rather sparse. It tends to show, however, that the growth of expenditure on research and experimental development in those countries is more rapid even when inflation rates are taken into account. But this growth is evidently small in absolute figures, so that the yearly effective growth of the research potential of the industrialized countries is very much larger than that of other countries. However, in matters of growth, only the relative increases decide the possibility of recovering the lost ground.

The disequilibrium between the industrialized third of humanity and the two-thirds on the road to industrialization, so far as their potential for innovation is concerned, could therefore be progressively diminished by a higher rate of growth of expenditure on science in the second group of countries. But since the growth is also very rapid in the first group, the effort that the developing countries will have to make in order to catch up is enormous. The success of this effort depends not only on the expansion of higher education, but also on measures which make it possible to give research scientists and engineers in their own countries the working conditions which they would otherwise be tempted to seek by emigration.

In this matter, as in that of economic growth, it is undoubtedly of the utmost importance that the rates of growth of the developing countries should be the highest. Here in fact lies the only hope of a convergence of these evolutions in the long run. The figures of Table 8 show that such an evolution is not out of the question.

TABLE 8. The evolution of the current expenditure on research and experimental development

Country	Date	Total of expenditure (in national currency)	Annual average growth during the period under consideration	National expenditure on R & D in percentages of the GNP
United States	1955	5.7 billion[1] dollars	13.9	1.4
	1966	23.8 billion dollars		3.0
U.S.S.R.	1959	2.8 billion roubles	9.0	2.1[2]
	1966	7.2 billion roubles		3.1[2]
United Kingdom	1955/56	0.25 billion pounds	11.2	1.3
	1964/65	0.65 billion pounds		1.95
Federal Republic of Germany	1956	2 billion marks	13.3	1.0
	1964	7.9 billion marks		1.9
France	1958	2.4 billion francs	21.1	0.97
	1965	9.2 billion francs		1.98
Czechoslovakia	1962	4.2 billion crowns	10.7	2.4[2]
	1966	6.3 billion crowns		3.3[2]
Belgium	1961	4.3 billion Belgian francs	11.7	0.72
	1965	6.7 billion Belgian francs		0.81
Japan	1961	177 billion yen	19.7	0.96
	1965	363 billion yen		1.19
India	1955/56	144 billion rupees	19.3	0.14[3]
	1965/66	841 billion rupees		0.42[3]
Republic of Korea	1963	1.33 billion won	29.1	0.28
	1966	2.86 billion won		0.31
Pakistan	1961/62	56 million rupees	44.5	0.15[4]
	1963/64	117 million rupees		0.27[4]
Philippines	1960	12 million pesos	36.0	0.095
	1964	41 million pesos		0.215
Cambodia	1963	1.0 million riels	100.0	0.004
	1965	4.0 million riels		0.013
Ceylon	1961	9.8 million rupees	6.0	0.15
	1965	12.4 million rupees		0.15

1. Billion = 1,000 million.
2. In percentage of the net material product.
3. In percentage of the net national product.
4. In percentage of the gross internal product.

Sources: Unesco and OECD.

EVOLUTION OVER A LONG PERIOD

The statistics and retrospective studies available do not yet make it possible to trace the evolution of national expenditure devoted to research and experimental development back over a long period except in the case of one single country, the United States of America. The graph below sheds light on the rapid growth of the scientific potential of this country since 1935. In fact, the share of the gross national product devoted to research and experimental development has increased between 1930 and 1965 from 0.2 to 3.0 per cent, that is to say, it is now fifteen times greater than it was thirty-five years ago.

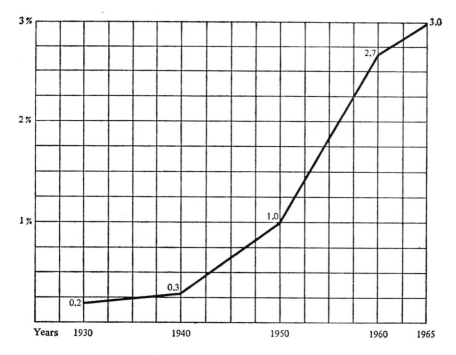

National expenditure on R & D in percentage of GNP (source: *Reviews of National Science Policy: United States*, OECD, 1968).

The curve is plainly exponential in its first part. However, since 1960 the expected slowing-down has begun, since on all the evidence the phenomenon follows a logistic curve and so tends towards a horizontal asymptote, though its level cannot yet be predicted with sufficient probability.

This graph shows vividly the 'explosive' character of the scientific expansion of our age. It is probable that every country in the world is following a similar curve at this moment—sometimes an even steeper one—but with a time-lag at the beginning.

Although it would be difficult to make a forecast from the fragmentary information available, it may well happen that, before the end of this century, the world will devote one-twentieth part of its economic resources to research and experimental development. The scientific and technological revolution would then be accomplished.

COMMENTS AND CONCLUSIONS

The scientific potential of the world is at present concentrated in the industrialized countries. At the heart of the group which they form, the centre of gravity of research is found in the most advanced countries, which involves a second stage of concentration of the world's potentialities of development.

As a result of accelerated technological progress, stemming from R & D efforts, the economy of the advanced countries at present appears to be growing as fast as that of the developing countries, and sometimes faster. At first sight therefore—and since the development process has become 'self-accelerating'—one might conclude that the less advanced countries will be kept definitely at a distance with no possibility of eventually catching up.

However, such a conclusion would be superficial. If, at a given date—in the present case 1965—the scientific potential of the nations is, in effect, more than proportional to the level of their gross national product per head, one finds that in 1930, at a time when their first industrialization was completed, the United States of America had a 'national R & D coefficient'[1] of 0.2 per cent, a figure found today in countries which have hardly begun the process of industrialization. Countries which have now reached a *per capita* gross national product similar to that of the United States of America in 1930, are spending many times more on research and experimental development than was spent by the United States of 1930. One might therefore conclude by analogy that the non-industrialized world today has a higher scientific development than that of the United States and Europe during the nineteenth century.

It is impossible to find, among the patterns of growth displayed by nations having different 'starting dates', the kind of resemblance that there would have been if exclusively endogenous factors were controlling the various stages of their development.[2] Therefore, with respect to all the countries of the world, the scientific explosion ought to be considered as an exogenous factor of development. This is a characteristic feature of the twentieth century. The response of the nations to this exogenous factor is strong, in absolute terms, when they have more resources available and when original technological innovation is the main factor of their growth. But the relative response of the developing nations is stronger because, when placed in the sphere of influence of modern science, these nations react more vigorously in propor-

1. The 'national R & D coefficient' is a country's total expenditure on research and experimental development expressed as a percentage of the gross national product.
2. Human beings, in the course of their embryonic life and their infancy, undergo a process of development which is essentially endogenous. They therefore remain similar in their growth whatever the starting date of the process. It is not the same with social organisms like nations, industries, etc. Exogenous factors can modify their evolution in a much more positive manner. The organic theory of social groups is clearly only valid within certain limits.

tion to their economic strength. The developing countries are not therefore suffering from a decisive loss of momentum in comparison with the advanced industrial countries.

However, it is absolutely essential for the developing countries that their rates of growth—of both gross national product per inhabitant and national R & D coefficient—should be higher than those of the industrialized countries; otherwise the evolution of the world would take an irreversibly disruptive course.

It is the joint responsibility of all States to avoid such an outcome.

5 Historical survey of the promotion of research and of government structures for science policy

The vital role played by governments in scientific affairs and the formulation of a specific science policy, at national level, are the results of initiatives taken during the last two or three centuries in a number of countries. We shall trace briefly their historical development, as well as the present status of the main government policy-making bodies in the field of science and technology.

PATRONAGE

The beginning of aid for research was the support afforded to men of science in the past by rich patrons. This kind of assistance prevailed roughly until the end of the seventeenth century. Most scientific work was dependent, in this age, on the goodwill of certain princes, who for their own prestige, or as a hobby, or out of their personal inclination, started museums of curios and provided an income for scholars (unless the researchers were themselves gentlefolk or wealthy bourgeois who found a use for their spare time and money in scientific inquiry).

For a long time the community of scholars suffered from the fact that this kind of support was inadequate to meet the needs—especially in terms of equipment—created by the development of experimental methods. Being anxious to exchange ideas, and to pool their funds in order to build certain expensive pieces of apparatus, they began to group themselves into clubs.

ACADEMIES AND LEARNED SOCIETIES

These clubs at first established themselves in academies. The first ones appeared in Italy: l'Accademia dei Lincei in Rome in 1600, and l'Accademia del Cimento in Florence in 1657. Their foundation was due to the direct initiative of two princes: Frederigo Cesi for the first, Leopoldo dei Medici for the second. These two institutions had a short life, since the Academy of Rome ceased the study of physics and astronomy with the condemnation of

86

Galileo in 1633, and that of Florence disappeared at the end of ten years, its members having for most of the time confined themselves to carrying out experiments, without however constructing any interpretations on the basis of those experiments; at that time it was less dangerous to collect the facts than to draw conclusions from them.

The two academies which followed had a very different fate and a very different reputation, since they were in some respects the model for all those which were founded later: these were the Royal Society of London and the Academy of Sciences at Paris, both founded in 1660.

The Royal Society, formed by a group of men of science and rich dilettantes, was recognized by King Charles II in 1662. The Academy of Sciences of Paris was instituted by royal decree on the initiative of Colbert. The latter—urged on by the example of Richelieu, who had inaugurated the Académie Française, and also by the opening of the Royal Society in London—acceded to the demands of the scholars who were grouped round certain personalities like Father Mersenne and Henri de Montmor, and who wished to obtain moral and financial support from the King's authority.

The academies of London and Paris were the forerunners of many institutions established on a similar pattern throughout Europe in the eighteenth and nineteenth centuries. Among the most famous may be mentioned: the Academy of Berlin, founded in 1700 through the untiring perseverance of Leibniz, the Academy of St. Petersburg in 1724, that of Stockholm in 1739, the Academy of Göttingen in 1751, and that of Munich in 1759. In 1772, on the proposal of Charles of Lorraine, the Empress Maria Theresa gave her Letters Patent to the Imperial and Royal Academy of the Sciences and Literature of Brussels. The Academy of Amsterdam was founded in 1808, that of Vienna in 1847, and the National Academy of Sciences of the United States of America in 1863.

Alongside these academies, whose aims were general, there were founded in several countries, especially at the end of the nineteenth century, scientific societies for separate disciplines: geology, botany, zoology, astronomy, geography. This was the beginning of specialization in science; it was partly due to the fact that the academies had begun to live in a closed circle, giving a too exclusive preference to certain branches of science, namely those on which the reputation of their leading members rested.

The foundation of the academies appears to some extent as a reaction against the monopoly exercised in those days by the universities over all intellectual activity. With a few exceptions the universities had in fact shown themselves to be somewhat disinclined, during the seventeenth and a great part of the eighteenth century, to organize a scientific curriculum based on observation and experiment rather than on accepted ideas. Moreover, the interests of the Royal Society and the Academy of Sciences were not exclusively confined to the pure sciences, but extended to more practical problems involving commerce, and the arts and crafts (today we would say the economy, and technology). Moreover, it was often in the exercise of this aspect of their activities that the sovereigns gave them their support. In the first years of the existence of the Academy of Sciences and the Royal Society, papers on the practical application of science were indeed more numerous than those concerned with the more fundamental themes of physics or mathematics.[1]

Up to the middle of the nineteenth century, the academies played a predominant part in support of the development of experimental science. They were a meeting-ground where men of science of all nationalities could compare their observations and their interpretations. The academies put at the disposal of scholars the resources which the latter would not have been able to muster individually. The academies also collected a mass of information on the condition of the sciences and the emergent techniques associated with them. Several of the academies assumed the role of advisers to the government on what decisions to take in scientific matters of national interest.

But although their appearance had given rise to undeniable progress in Europe on the plane of the encouragement of the sciences, nevertheless two factors limited the scope of their activity when science was confronted with the great industrial revolution. These factors were: lack of money, and the highly independent character of the research carried out by the academicians.

With the exception of the Academy of Paris which, being largely financed by the public treasury, was almost a Department of State, they all lacked the necessary funds to carry out a real advance in scientific research. Besides, being limited to the simple provision of the means for research and to the discussion of individual operations, they never formed proper research teams, such as scientific progress demanded.

In the majority of Western countries, the actual role of the academies consisted above all in putting their libraries at the disposal of the men of science, in publishing reports, reviews and memoranda, in organizing conferences and lectures, and in giving prizes.

In general, the learned societies of the Western world were 'more ready to reward than to undertake'.[2] They confined themselves to an advisory role, but often in a second place or in duplication with other organizations.

For the Soviet Union, on the other hand, the Revolution radically changed the role of the academy, which lost much of its independence from the State (up to the war of 1914, it had often openly opposed the government, especially after the first abortive revolution of 1905). It ceased to be a learned

1. Sir Henry Lyons, *The Royal Society 1660-1940*, p. 114, 131, 154, Cambridge, CUP, 1944; James E. King, *Science and Rationalism in the Government of Louis XIV*, p. 294, 608, Baltimore, Johns Hopkins Press, 1949.
2. The Royal Society of London, however, has fulfilled a larger function by bringing out yearly a review and critical analysis of the research work done in the country. For a number of years it supplied the Advisory Council on Scientific Policy with reports on the areas of research which are insufficiently covered. In the United States, the Academy of Sciences has intervened in a still more direct manner to advise the Federal Government by inaugurating, during the First World War, the National Research Council which was entrusted by the government with the task of reinforcing the scientific efforts of the country to fulfil military needs. However, it lost its role of principal adviser to the government after the creation, in 1950, of the National Science Foundation. Turning more to Congress, the Academy succeeded in getting itself regarded as official adviser to this body and became thereby a sort of counterbalance to the advisory councils of the executive branch of government.

society and a prize-giving institution only, and became instead an organ of State, taking part directly in the administration of the country: a sort of central organization for the management and execution of scientific research. Most Communist countries followed this example.

INSTITUTIONS FOR THE PROMOTION OF RESEARCH

From the end of the nineteenth century onwards, the development of industrial capitalism favoured the development of a particular form of disinterested aid afforded to research scientists: the philanthropic foundation.

In certain countries, private foundations constituted—and still constitute today—a relatively important source of finance of scientific research by granting prizes and scholarships, and by allocating aid to individual research workers or to scientific institutions. In certain cases they themselves organize research activities.

In Europe however private foundations never assumed the importance which they enjoyed in the United States of America, because the fiscal legislation of the European countries was usually much less generous in the matter of the tax deductibility of donations made in favour of philanthropic institutions.

At the present time more than two hundred American foundations are still financing scientific research.[1] Their expenditure on research amounts to some eighty million dollars; half of this total is taken up by medicine and biology. In the Federal Republic of Germany, the foundations[2] and certain private societies for the encouragement of research, which can be compared to foundations (Deutsche Forschungsgemeinschaft, Max-Planck-Gesellschaft)[3] still constitute important instruments for the financing of scientific research; by the intermediary of these societies, industry can give support to research outside the framework of its immediate professional interests.

On the other hand, this kind of institution has not been extensively developed in France or the United Kingdom.

Although—thanks to their statutory functions and the kind of assignment imposed on them by their benefactors—the scientific foundations were able to develop an efficient apparatus for responding to the initiative of research, they generally showed themselves less suited to the formulation of national development goals and specific objectives for research, or to the planning of research programmes. Moreover, the income of these foundations rapidly became inadequate in comparison with the rising magnitude of research needs. Governments were therefore increasingly led into financing important research programmes.

1. The largest are the Carnegie Institution (1902), the Rockefeller Foundation (1913), and the Ford Foundation (1936).
2. The leading foundations in the Federal Republic of Germany are: the Stifterverband für die Deutsche Wissenschaft (1949), the Fritz Thyssen Stiftung (1960), the Stiftung Volkswagenwerk (1961), and the Alexander von Humboldt Stiftung.
3. The last-mentioned, already financed up to 90 per cent by public funds, is in the process of evolution towards a *de facto* governmental status. The case of the National Fund for Scientific Research in Belgium is analogous.

The United States of America is virtually the only country where, because of the deductions agreed to by the Treasury on income tax returns, foundations actually receive important contributions from private companies for the promotion of scientific activities.

CHARACTERISTICS THAT THE ACADEMIES AND FOUNDATIONS HAVE IN COMMON: THE CRITERION OF MERIT

The majority of the academies and foundations did not succeed in adapting themselves to contemporary needs. While this was certainly due to their inability to collect the necessary financial resources, it was also the result of a too exclusive criterion which guided their policy: research was only supported according to the merit of the applicant scientist, since the value of the research projects was assessed by the scientific community itself.

While this principle still remains largely valid for the support of basic research and for those projects where cost is limited, it has become inadequate when what is at stake is the financing and advancement of research activities directed towards practical applications, or very expensive fundamental research involving projects such as particle accelerators or observer satellites. In such cases it is necessarily on the ground of specific objectives that the support of research is granted; considerations of personal merit are no longer sufficient and have to be supplemented by other criteria, such as social utility, economic profitability, and efficiency in fulfilling the objectives set forth for the research. The objectives of the research and the quality of the management become dominant criteria—but the academies and the foundations are usually not empowered to examine research projects from this point of view.

THE GOVERNMENTS ENTER THE FIELD

The inadequacy of private initiatives induced the governments to consider much more far-reaching measures to promote research.

SCIENTIFIC INSTITUTIONS OF THE GOVERNMENT

In the course of time, governments themselves organized research activities in many spheres of public interest which were left to them even in the most *laissez-faire* capitalistic societies: collection of meteorological and astronomical data, geological surveys and topographic mapping, surveys of flora and fauna, public health, the determination of standards for food and drugs, improvement of agricultural production, improvement of transport and communications, etc.

In this way a network of government establishments,[1] discharging an office

1. Some of these establishments were of ancient origin, such as the Royal Botanical Garden of medicinal plants in Paris, which became the Museum of Natural History, the Observatory of Paris (1667) and that of Greenwich (1675), the Central Astro-

of scientific public service, developed in the countries of Western Europe and in the United States of America.

THE FINANCING OF UNIVERSITY RESEARCH OUT OF PUBLIC FUNDS

In the majority of countries, private initiative showed itself inadequate—besides being inadequately motivated—to cope with the huge needs of education; governments therefore took over responsibility for the bulk of expenditure on education at all levels, and especially for the regular budgets of the universities. In this way they assumed, in a general manner, direct responsibility for the training of administrative and scientific cadres and for the support of fundamental research. This responsibility still falls today on national governments, whatever the status of the universities in other respects, and many indeed maintain their claim to autonomous or private status.

But in the majority of countries the role of the government in its dealings with the universities was confined to determining the conditions for the conferment of academic degrees, and to financing their activities by means of block grants. Complete liberty was generally left to the universities as to the detailed assignment of the funds received and their distribution between educational and research purposes.

GOVERNMENT ORGANIZATIONS FOR THE PROMOTION OF RESEARCH

Government support of scientific research projects outside the governmental establishments (*extra-muros* research) remained spasmodic until the First World War, and was limited in practice to the financing of certain exceptional enterprises such as scientific expeditions.

The first important measures of public support for research projects were taken in the United Kingdom and in France during the 1914-18 war and immediately after it.

Military operations showed the larger belligerent nations that science could put a trump card in their hands. France, the United Kingdom and even the United States of America, faced with the problem of the mass production of increasingly sophisticated armaments, realized how far their industries had fallen behind German industries, which turned to better account the discoveries of science.[1] These countries found themselves obliged to set up organizational structures whose aim was to mobilize the scientific resources necessary to catch up and overtake.

nomical Office in Paris, founded in 1675 for the use of the Navy and commerce. One can also quote the laboratories of the Physikalisch-Technische Reichsanstalt in Germany, founded in 1880, and in England the National Physical Laboratory, founded in 1900. In the United States of America the geodesic and hydrographic service, and the laboratories of the Department of Agriculture, played an important part, especially in the prospecting and commercial development of the western territories, at the end of the nineteenth and the beginning of the twentieth centuries.

1. During the First World War also, the idea—already foreseen by some classical economists—took root that research, by conditioning the progress of industry, itself constitutes a factor in economic growth.

In various countries governments chose either to create public organizations of a new kind for this purpose, or else to rely on existing private institutions, whose financing they gradually took over. Both these forms of intervention ended up, in fact, with the same result. The distribution of the sums allotted by the government for research were entrusted to central organizations for the promotion of research, which were largely autonomous and operated under the managerial guidance of committees composed of scientists.

The way these organizations actually operated varied according to the country. In certain countries they confined themselves to giving fellowships, grants and subventions for research, either to individual research workers or to research teams; in others they also operated, at the same time, their own research laboratories. By creating their own research institutes, often grouped around large laboratories and apparatus, these organizations compensated the deficiencies of the universities.

The *United Kingdom* was the first to develop a stable and efficient government organization to promote scientific activities. In accordance with the recommendations of the Haldane Commission, a body endowed with a large measure of administrative and financial independence—the Department of Scientific and Industrial Research (DSIR)—was created in 1915.

The task of the DSIR developed over the years on two planes: on the first was general promotion of fundamental research, and on the second, encouragement to apply the results of research to industrial development. The Department gave financial support to university research workers, and grants to post-graduate students; it built and financed its own laboratories, and financed industrial research associations, etc. The management of the Department was entrusted to a Research Council composed of representatives of the scientific and industrial communities.

Other organizations were subsequently created following a similar administrative pattern: the Medical Research Council in 1920 (which replaced the Medical Research Committee created in 1911); the Agricultural Research Council, in 1931; the Nature Conservancy in 1949; the Overseas Research Council in 1959, etc.

Until recently these different institutions presented their budget directly to Parliament. This pattern of organization has been modified recently: the powers exercised by the DSIR were split up between a Science Research Council, placed under the control of the Department of Education and Science, with authority to support university research, and a Ministry of Technology under whose control came all the activities of the DSIR in the industrial field.

In *France,* before the Second World War, various attempts were made, but without success, to promote scientific research. To this end there were created: in 1868, the École Pratique des Hautes Études, and in 1905, the National Science Fund, which became, in 1935, the National Fund for Scientific Research.

In the field of applied research, a Department for Inventions Relating to National Defence was founded in 1915. The Department became, in 1922, the National Office for Scientific Research and Inventions. Its activities rapidly proved insufficient. In 1938 the threat of war forced the government to create a National Centre for Applied Scientific Research, which replaced the Office.

None of these institutions had sufficient funds at their disposal.

A short time after the beginning of the Second World War, the National Centre for Scientific Research (CNRS) was founded, as a result of the merger of the National Fund for Scientific Research and the National Centre for Applied Scientific Research. The CNRS, a government establishment placed under the authority of the Minister of Education, received in principle a very general task—to develop, direct and co-ordinate scientific research of all kinds—which made it the central organization in the promotion of research. In fact, however, its initiatives were directed almost entirely towards university research. To fulfil its mandate, the CNRS had the powers to carry out research in its own laboratories, to finance individuals or institutions and to organize post-graduate courses of training in research.

In the *United States of America,* an Office of Scientific Research and Development (OSRD) was founded in 1940, to serve as a centre for the mobilization of the scientific personnel and resources of the nation, with the object of ensuring their best use for developing the results of scientific research and applying them to the defence of the country. It was attached to the Office of the President of the United States. The OSRD did not possess laboratories of its own. Its interventions consisted in financial help given to research workers and to research centres. The Office disappeared at the end of the war, but the need persisted. Discussions lasted for five years, and it was not until 1950 that the National Science Foundation was inaugurated. Its task was to finance fundamental research and higher education by means of grants and fellowships.

In the *Netherlands,* two central government organizations are in operation: the Nederlandse Centrale Organisatie voor Toegepast Natuurwetenschappelijk Onderzoek (TNO), created in 1932 for the promotion of applied research carried out in the public interest, and the Nederlandse Organisatie voor Zuiver Wetenschappelijk Onderzoek (ZWO), created in 1950 for the promotion of fundamental research.

In *Sweden,* a certain number of research councils were established by the government between 1940 and 1950 to distribute the funds which the State devoted to the promotion of research. Similar organization was established in Norway.

In *Belgium,* after a first attempt in 1921 had failed to develop for financial reasons, a central institution for the promotion of research was founded in 1928, but at the beginning the formula adopted was that of a private foundation: the National Fund for Scientific Research (FNRS) which was set up utilizing contributions from patrons and subscription by the large banks and industries.

At the beginning, the sphere of intervention of the FNRS covered fundamental research as well as applied research. However, the means at its disposal were insufficient to conduct an efficient operation in support of research in industry. The State therefore also created the Institute for the Encouragement of Scientific Research in Industry and Agriculture (IRSIA). Now the

FNRS itself operates largely by means of public funds. However, it has kept its private status.

GOVERNMENT CONTRACTS FOR RESEARCH AND EXPERIMENTAL DEVELOPMENT

In every country, but to a different extent, the government has direct recourse to outside organizations—industries, universities, independent laboratories—to carry out specific pieces of research which are of importance to the nation or to find the answer to questions which confront the State in the course of its administration.

Since the Second World War, the action of governments under this heading has taken on an entirely new dimension; several countries have embarked on huge military, space and nuclear programmes, which make very large demands for scientific and technological research. Development of these government initiatives, in the scientific sectors involved in carrying out these programmes, required very different procedures from those which until then had been considered suitable for the promotion of research, since these more recent initiatives aimed at urgent practical applications, and demanded the use and control of much larger financial resources.

The administration of these programmes is usually in the hands of specialized government organizations, which are administratively autonomous and qualified to cover a large research area. These organizations usually possess their own laboratories, but also entrust the carrying out of research to outside organizations—industrial and university. The volume of this kind of research often far surpasses that which they carry on *intra muros*.

One therefore witnessed the creation, in the United States of America, of the National Aeronautics and Space Administration (NASA) and the Atomic Energy Commission (AEC); in the United Kingdom, the Atomic Energy Authority (UKAEA); in France the Commissariat à l'Énergie Atomique (CEA) and the Centre National d'Études Spatiales (CNES).

This last form of intervention by governments has given rise to elaborate contract systems linking, on the one hand, a government and its subsidiary organs, and on the other hand private enterprises, research teams and institutions, such as the universities, which are independent of the State. In several countries, these contracts have profoundly modified the relations between the scientific community, industry and government. They have directed a nation's scientific potential into particular channels, strictly corresponding to national goals and to the objectives of the government.

This policy of government contracts has assumed an enormous importance in the United States, as much for the size of the funds that it puts at the disposal of research as for the organizational changes which it has set in motion. The most clear-sighted observers of American society see in it one of the causes of the great change which has taken place, over the last twenty years, in the relations between industry and government, and the emergence of a new kind of society.[1]

1. See John K. Galbraith, *The New Industrial State,* Boston, Mass., Houghton Mifflin Co., 1967.

THE APPEARANCE OF GOVERNMENT STRUCTURES FOR SCIENCE POLICY[1]

Government interventions on behalf of science ended up by composing a somewhat incongruous aggregate, each element of which, whether procedure or institution, provided a response to a specific need which was felt at a given moment.

So long as the entire expenditure needed in this situation was of modest proportions, compared with the total government expenditure, it did not attract enough attention to justify measures of co-ordination. We have seen that attempts were made in this direction in the nineteenth century and at the beginning of the twentieth century. They were obviously premature.

When scientific expenditure began to represent a significant item in the budget, governments were forced to regard their dealings with science from a new angle: that of the general policy of the State. But a definite level had to be reached for the political authorities to become aware of the need.

UNITED KINGDOM

The United Kingdom was the first country to establish a genuine central organization for the promotion of research, and to devise a governmental pattern for the control of its national science policy.

This pattern aimed at fulfilling two essential functions: first, that of co-ordinating the powers exercised by the different ministerial departments on behalf of research; and secondly, that of formulating a national science policy which would satisfy all the interested parties—the government, the universities and industry.

The principles of the British system had been partly foreseen as far back as 1871, when one of the recommendations of a Royal Commission had been to suggest a Minister of State responsible for all activities in the sphere of science, who should have an independent consultative body to assist him, and to install similar consultative bodies in the War Office and the Admiralty.

In fact this organization was only adopted in 1947 when the Advisory Council on Scientific Policy and an advisory council for defence research took the place of the war-time commissions. These two committees were at first presided over by the same person.

In 1959 a Minister for Science was nominated, who took over the duties exercised in the scientific field by the Lord President of the Council,[2] and

1. In this section, considerations of space have made it necessary to reduce to the dimension of mere 'thumb-nail sketches' the account given of the evolution and current situation of the national science policy structures of certain of Unesco's Member States. For a fuller and more recent review of the situation in twenty-six of Unesco's European Member States, including detailed organizational charts, the reader is referred to *National Science Policies in Europe*, no. 17 in the series 'Science Policy Studies and Documents' (Unesco, Paris, 1970).
2. In 1916 a committee of the Privy Council, presided over by the Lord President of the Council, was entrusted with the task of promoting scientific research. Thus the Lord President, a kind of minister without portfolio, became responsible before Parliament for the expenditure of the Department of Scientific and Industrial Research (DSIR).

this time the Minister was put at the head of a department. Various ministers, such as the Ministers of Aviation, Defence and Power, nevertheless kept exclusive control over important research activities. The co-ordination of the whole of government expenditure on science was therefore not yet achieved.

A commission established by the Macmillan Government submitted a report (known as the 'Trend Report') in 1963, with recommendations which aimed basically at certain changes of organization, with the object of defining more distinctly responsibilities for aid to industrial research and aid to fundamental research and, with that end in view, to split up the powers of the DSIR.

The commission also proposed to combine the Ministry of Education and the Ministry of Science. This last recommendation was put into effect by the next government; thus a Department of Education and Science came into being, with control over education at all levels. In addition to the functions which had been exercised by the previous Minister of Science, it found itself entrusted with the control of scientific activities which had formerly been the province of various ministries. The budgetary responsibility of the Minister became heavier. More particularly it extended to the Research Councils, which previously presented their budget directly to Parliament. The Minister was at the same time entrusted with the supervision of the activity of the Councils.

The reforms thus begun were completed by the creation, simultaneously with the Department of Education and Science, of a Ministry of Technology, responsible at once for the activities of the United Kingdom Atomic Energy Authority, for those industrial research activities which had previously been the responsibility of the DSIR, and for the placing of research contracts in industry. The Ministry of Aviation was absorbed, but defence research was not put under the control of the new ministry.

As the situation stands, three ministers still have a hand in British research: the Minister of Education and Science deals with fundamental research and research in universities, the Minister of Technology[1] deals with industrial research, and the Minister of Defence deals with research for military purposes.

The existence of such a three-headed structure has made necessary the creation of an advisory council charged with the task of advising the government on the over-all aspects of its science policy. This is the Central Advisory Council for Science and Technology (CACST) which operates under the chairmanship of the Chief Scientific Adviser to the Government.

In a sense, therefore, the CACST is more centrally situated than any one of the individual ministries concerned with science, for its terms of reference require it to report to the Prime Minister on 'strategic' matters concerning the country's science and technology resources as a whole. It has to be recognized, however, that its members—chosen as representing many fields of national activity—serve on an unpaid, part-time basis and in an advisory capacity only. Moreover, the number of officials acting as the secretariat of

---

1. Now the Department of Trade and Industry.

the CACST is very small, and these officials are not employed exclusively on CACST affairs.[1]

It may therefore be doubted whether, although the United Kingdom took the initiative in nation-wide scientific co-ordination, it has yet achieved—despite several changes of organization in its science policy—a unified decision-making system in this field. It also looks as if the countries which have copied the British organization are finding themselves in the same situation.

The road towards a ministry of science, which would regroup all the scientific departments of the government, is seen to be paved with formidable obstacles. It has proved very difficult in practice to end up with less than two ministries of science and to take away military research from the Ministry of National Defence. Must it then be concluded that co-ordination by re-grouping ministerial responsibilities in the field of research and experimental development is a dead end?

FRANCE

The origin of the attitude of the French Government authorities towards science must no doubt be looked for in the traditional tendencies of the State towards centralization.

But during more than a century the attempts made to establish a machinery capable of adopting a complete and consistent science policy remained, as in the United Kingdom, at the stage of intentions.

A first step was made by the creation, in 1933, of a Higher Council for Scientific Research, whose mandate was to tender advice and proposals to the Minister of National Education, and to operate the distribution of the resources of the National Science Fund. The office of Under-Secretary of State for Scientific Research was created in 1936. The first holder was Irène Joliot-Curie, but the post was abolished at the end of 1937.

After the war, in 1945, the government entrusted the National Centre for Scientific Research (CNRS) with the mandate of 'ensuring the co-ordination of the research carried out by the government departments, industry and individuals'. To enable it to fulfil this assignment, the CNRS was empowered 'to carry out a survey in government and in private laboratories on the research which they pursue and the funds which they have at their disposal'. However, the Centre very quickly ran up against difficulties which it did not succeed in overcoming. As matters stood, a large number of research activities had been undertaken in other ministries, besides that of national educa-

---

1. Administratively speaking, these officials—who are of course high-ranking Civil Servants—come under the authority of the Chief Scientific Adviser to the Government, and are part of the staff of the Cabinet Office.

    This Office also provides the secretariat for a number of standing, and also *ad hoc*, interdepartmental 'Cabinet Committees'.

    The day-to-day co-ordination of viewpoints and interests in science and technology, as between the various departments concerned, depends in practice as much on this network of 'Cabinet Committees' as on the functioning of the CACST.

tion. The CNRS, put under the wing of the Department of Education and itself engaged in carrying out projects of research, found itself ill equipped to make this survey. Further, it did not enjoy the necessary authority to ensure a co-ordination of research, for this required an authority—which it did not possess—over the different ministries engaged in the operations of conducting and financing research.

Meanwhile, various projects were conceived; they aimed at establishing a more efficient organization for co-ordination. Their successive appearance and disappearance were not entirely due to the instability of the government of the day, but above all to a very real concern to improve the way in which the organization was set up, and to define its responsibilities more exactly, as the ideas on how to effect real co-ordination became more clear.

Thus, in 1947, a Bill proposing the creation of a Higher Council for Research was presented to the National Assembly. This Bill for the first time put in a concrete form three essential conceptions: (a) policy guidance to be given to the research organizations, (b) a national survey of the existing R & D means and resources, (c) the 'over-all plans for research'. It did not however succeed, because it continued to base the co-ordination of scientific activities on the Ministry of Education alone.

In 1953, the Economic and Social Council recommended, first, to entrust the responsibility of promoting research to the head of the government, who would delegate his powers in this field to a Secretary of State, and secondly to create a small consultative council.

This project was founded on two principles: the responsibilities for co-ordination ought to be placed at the highest level, under the Prime Minister himself, and the task of putting forward a policy ought to be entrusted to a widely representative consultative council. A third principle was soon to be added: namely that of the integration of the national science policy with the Plan for Economic and Social Development. In 1953, a Commission for Scientific and Technical Research was created, within the framework of the 'Commissariat Général au Plan'.

In 1958 the government created, under the authority of the Prime Minister, an Interministerial Committee of Scientific Research, and a Consultative Committee of Scientific and Technical Research.

The Interministerial Committee brings together the most important ministers interested in scientific questions. It forms therefore the organ of inter-departmental co-ordination towards which all efforts had been aiming since the CNRS was no longer felt adequate for this task. To prepare the agenda for the Interministerial Committee, and to help it in its work, the president of the committee co-opts, for a maximum period of two years, individuals chosen for their qualifications in matters of research or economics. They sit in an advisory capacity with no voting right. These individuals in fact comprise a consultative committee[1] inside the Interministerial Committee.

A common secretariat—the Délégation Générale à la Recherche Scientifique et Technique (DGRST), answerable directly to the Prime Minister—serves both committees. It is directed by a senior civil servant, the General

1. *Report on the Organization of Scientific Research in France*, Paris, OECD, 1964.

Delegate, who is empowered to set up working parties for particular subjects, and to co-opt members with particular qualifications. The mandate of the General Delegation is: (a) to take action on all reports, studies and inquiries concerning the situation and development of research, and to make a survey of the human and material resources of research; (b) to collect the budgetary proposals of the different ministerial departments involved, and to ensure their examination with the aim of presenting them for discussion at the Consultative Committee and the Interministerial Committee.

In addition, the General Delegation prepares the work of the Commission for Scientific and Technical Research of the 'Commissariat au Plan'.[1] A significant point remains: the organization just described, and more particularly the powers of the General Delegation, are concerned only with the civil aspects of research and, of such aspects, exclude those connected with the nuclear or space fields.

In these two last fields, special independent organizations, directly answerable to the Prime Minister, have been entrusted with both the co-ordination and the execution of research: the 'Commissariat à l'Énergie Atomique' (CEA) and the 'Centre National d'Études Spatiales' (CNES).

In the military sphere, the co-ordination of research has been entrusted since 1959 to a consultative body, under the aegis of the Prime Minister, the 'Comité d'Action Scientifique de Défense', which has powers similar to those of the Consultative Committee in the civil sphere.

However, the powers which relate as much to scientific research as to atomic and space matters have been grouped in the hands of one individual member of the government, who thus, in his own person, brings about a kind of union between the independent administrative and consultative organizations. Despite this union through one individual, the responsibility is still, in fact, shared between at least two ministers, the Minister for Scientific Research and Atomic and Space Affairs, and the Minister of War.[2]

Above all it will be remembered that an essential characteristic of the French system was the determination to conceive the scientific plan in close relationship with the economic and social plan.

UNITED STATES OF AMERICA

As in the United Kingdom and France, the idea of creating government organs for science policy dates back to the last century, but it did not take shape until after the last war.[3]

Before the First World War, various proposals had been made in vain by the National Academy of Sciences, with the aim of co-ordinating the scientific services of the government.

1. Decree of 6 May 1953.
2. The situation has been changed again by the Decree of 11 August 1970.
3. For the whole of this section, it is useful to refer (a) to the study published in 1968 by Unesco, entitled *National Science Policies of the U.S.A.* (no. 10 in the series 'Science Policy Studies and Documents'); and (b) to *Reviews of National Science Policy, United States*, Paris, OECD, 1968.

During that war, at the demand of President Wilson, the Academy founded the National Research Council in order that it should tender scientific advice on problems of public interest, when required by the Federal Government. However, its influence in this respect ceased at the same time as hostilities.

Another short-lived body was created by President Roosevelt, namely the Science Advisory Board. It came into collision with the desire to preserve the activities of the scientists from the interference of the government.

Centralization under the direct control of the President, which it had been possible to realize during the Second World War, through the Office of Scientific Research and Development mentioned above, was no longer accepted in times of peace, and the OSRD was dissolved. Its director, Vannevar Bush, then recommended the creation of three organs:

1. A committee for the co-ordination of the scientific activities of government departments (which was created in 1947).
2. A consultative committee composed of scientists (which was established in 1950 within the framework of the Office of Mobilization for Defence, under the name of the Science Advisory Committee, though it was never very active and did not receive much attention).
3. A government institution for the encouragement of the fundamental sciences. (This only saw the light of day in the shape of the National Science Foundation in 1950, because of the numerous disputes which took place about its independence.[1] Once it was created, the NSF encountered similar difficulties to those of the French CNRS. Not having the right of inspection over the other scientific agencies of the government, it did not succeed in carrying out its mandate of giving 'an appreciation of the programmes of research undertaken by the agencies of the federal government'.)

On all the evidence, it appears that the event which administered the psychological shock necessary for the United States to speed up establishment of a proper organization of scientific policy, was the launching of the first *sputnik* in 1957. In face of the effect of the Soviet performance on public opinion and on the Members of Congress and government, President Eisenhower immediately took as an adviser a Special Assistant for Science and Technology and, at the same time, transformed the Science Advisory Committee. He took from it its exclusively military character, and linked it directly with the Presidency of the United States of America.

The Special Assistant is much more than a simple scientific adviser. Being directly in touch with the White House, he can supervise all the activities of departments. It is his business to keep himself informed of all the scientific activities of different departments, and to give personal advice to the President on the future direction of government policy in scientific matters.

The President's Science Advisory Committee is composed of representatives of the universities and of industry. It has been given a twofold mandate

1. See: Don K. Price, *Government and Science: their Dynamic Relation in American Democracy,* p. 48-55, New York, OUP, 1962.

of study and advice, bearing on the action of the government in favour of science and the integration of this action into the whole national effort.

In 1959, on a recommendation from the Science Advisory Committee, the President transformed the interdepartmental committee into a Federal Council for Science and Technology, whose mandate was to prepare and work out the co-ordinated planning of certain national programmes (the evaluation of needs and of scientific potential, the distribution of duties between the different government organizations, and the working out of the budgets). Various committees were allotted these tasks, and also that of making a study of the problems of administration of the federal programmes, and of the working out of a science and technology policy at governmental level.

The Special Assistant and the Advisory Committee shared at the beginning the same administrative and professional staff. However, it rapidly became clear that it was necessary to ensure a firmer position for this personnel in the government organization. Besides this, the Special Assistant, being directly linked to the White House, escaped the Congress's powers of inquiry. For these two reasons, President Kennedy created a new organization, the Office of Science and Technology, at the highest level of the administrative hierarchy, that is to say at the level of the Executive Offices of the President (like, for example, the Bureau of the Budget). Thus the Office comprises in fact an organ for the administration of science, having access to all information about the scientific activities of the government, and reporting directly to the President.

Liaison between the three science policy organs is ensured by the Special Assistant, who directs the Office of Science and Technology and presides over the two councils.

The characteristic of the American system is therefore the centralization of the scientific organization under the direct authority of the President,[1] and the entrusting to senior civil servants, answerable to the President, of the following tasks: the formation of science policy, the fixing of the budgets, the organization of interdepartmental co-ordination, and also consultation with the scientific community.

FEDERAL REPUBLIC OF GERMANY

In the Federal Republic of Germany, much more than in the United States of America, the way in which the State itself is organized has conditioned the nature, and the powers, of the science policy institutions.

The Constitution of 1949 in principle gave the Länder—the States of the Federation—legislative and administrative power in cultural matters, especially matters concerned with higher education (the organization and financing of the universities). By virtue of this same Constitution, the Federal Government only has parallel powers in legislation concerning the promotion of scientific research; that is to say the Länder may legislate in this

1. It must be noted here that even the NSF reports to the President of the United States of America, who appoints its director.

field whenever the Federal Government has not yet made use of its own legislative power. Further, various initiatives have been taken by the Länder since 1949, to achieve a better co-ordination between them (a standing conference of the ministers of education of the various Länder).

At the federal level, the co-ordination of all the scientific activities carried out by the various ministries of the Federal Government has been achieved since 1962 by an Interministerial Committee of Science and Research. The presidency of the committee was at first assigned to the Federal Minister of the Interior, but since 1963 has been exercised by the Federal Minister for Scientific Research[1] who, in addition to the scientific activities which he has taken over from the Minister of the Interior, has jurisdiction over all scientific questions as well as others which concern nuclear energy.

To ensure co-ordination between the policy of the Länder and that of the Federal Government, a Science Council (Wissenschaftsrat) was created in 1957. This institution contains two committees: one administrative, composed of senior civil servants of the Federal Government and the governments of the Länder, the other scientific, composed of representatives of the scientific and economic communities. The mandate of the Council was defined as follows: (a) to work out—on the basis of the plans made by the Federal Government and the Länder, within the framework of their respective powers —an over-all plan for the promotion of the sciences, and at the same time to bring the plans of the Federal Government and the Länder into line with each other; (b) to fix a programme of priorities every year; and (c) to make recommendations for the allotment of government grants, to be earmarked for research, by the Federal Government and the Länder.

BELGIUM

Belgium shares with some other countries the advantage of having grasped comparatively early the usefulness of deciding on an over-all policy for scientific development.[2]

The decisive step was marked by the creation in 1957, on the initiative of the government, of a National Commission for the study of the problems posed by the progress of science and their economic and social repercussions.

At the end of its deliberations, which lasted two years, the Commission submitted a report to the government with recommendations concerning the creation of a government structure for science policy. Three organs were created in 1959. These were:

1. The Ministerial Science Policy Committee (CMPS), which comprised all the ministers with powers in scientific matters, and also the Minister of Finance.

---

1. Bundesminister für wissenschaftliche Forschung. Since 1969, the responsibilities of this Minister have been enlarged: he is now Federal Minister for Education and Research (Bundesminister für Bildung und Wissenschaft).
2. For the Belgian organization it is useful to consult the Unesco report entitled *La Politique Scientifique et l'Organisation de la Recherche Scientifique en Belgique* (Paris, 1965), and the report of OECD: *Reviews of National Science Policy: Belgium* (Paris, 1966).

2. The Interministerial Science Policy Commission (CIPS), which groups together the senior civil servants responsible for scientific policy in the departments represented on the committee.
3. The National Science Policy Council (CNPS), which brings together the representatives of the scientific community, the universities, the economic world and the trade unions.

The three organs were put under the authority of the Prime Minister. Their powers extend equally over higher education and over scientific research, both fundamental and applied, including defence research (though this is of less importance in Belgium).

The tasks of studying and preparing national programmes have been taken over by a small multidisciplinary professional staff, which is part of the Prime Minister's department, and has recently taken the name of the Department for the Programming of Science Policy. This staff also constitutes the secretariat of the National Science Policy Council (CNPS). Its Secretary General is the Chairman of the Interministerial Science Policy Commission.

Part three

A policy of development based on science

6 Science policy as part of the general policy of the nation

INTRODUCTION

In Chapter 4 we stressed the vast differences which exist between scientific potentials and the government structures for science policy relevant to the four most typical stages of national economic development.

Some of these differences can be explained by the nature of the problems with which the peoples and their governments are actually faced. For some, agriculture is essential, while for others it has ceased to be so; for some, advanced technology provides the only means capable of promoting growth and taking up the slack of unemployment, whereas others have first-generation industries which must be built up as a top priority in order to free their balance of payments from the growing weight of the import of intermediate products or consumer goods, etc.

Yet one cannot help being struck by the considerable degree of failure to adapt scientific potentials to the real needs of the nations. The number of those engaged in research, and the expenditure on research itself, is inadequate; so that the countries which manifestly stand in greatest need of the help which science could provide, actually use it least. Paradoxically, the scientific development of nations appears to be the *consequence* of their economic development rather than its cause. This makes research look like a luxury—to be afforded only by those who have already overcome the problems of economic poverty. The apparent economic uselessness of space research, and the aristocratic attitude of some 'pure' scientists in the least developed countries, may give weight to this altruistic idea of science—an idea which, moreover, has deep roots in the tradition of Europe itself. It is, in our experience, needless to underline the plausible fallacy that it contains, since the technological successes of the Western world are clearly the foundation of its economic advance, and these technological successes have their origin in the utilization of the scientific approach and in the application of scientific knowledge to solve practical problems.

True, the scientific organization of nations more often reflects the needs of the preceding epoch than that of the stage of development which they have actually reached. Such organization hardly ever leads on naturally to the needs of a future stage, as it ought to do. Research being essentially a preparation for the future, the state of scientific organization of a country ought in fact to give some indication of the shape it will assume tomorrow.

The sad fact is, however, that there are only too many countries which, though they have made satisfactory progress in their first phase of industrialization, still have hardly any institutions for agronomic and industrial research, whether supported by the State or by private corporations. These countries remain content with a rudimentary infrastructure of public services, and their universities are still oriented chiefly towards the humanities and medicine. To sum up, their scientific organization has remained mediaeval even though their economy is almost modern.

In the same way, one sees European countries, already engaged in the second phase of their industrialization, which as a matter of tradition continue to apply their principal efforts to classical agronomy and industrial technology, and thus abandon to foreign influence key sectors for their present-day economic growth.

Failure to adapt scientific organization and budgets to the real needs of the nation has indeed been widespread. Out of this situation has emerged the growing interest in science policy, witnessed in all countries since the beginning of the century. This is also why science policy has often been envisaged at the beginning as a *policy for science*—that is, a series of measures which aim to remedy the chief deficiencies which have accumulated in the past, deficiencies which had prevented certain sections of the nation's research and development plant from developing normally. On the eve of the Second World War no other doctrine of science policy had, to our knowledge, acquired any importance.

As we have recalled in the previous chapter, certain countries had indeed taken note, in the period following the 1914-18 war, of the military dangers brought about by an industry that was behind the times in the applications of scientific knowledge. For their part the great enterprises, especially those in the sector of chemistry, were already well past the stage of a vague awareness of scientific backwardness, and were beginning to put in hand specific research programmes with a view to the attainment of precise objectives. But the *strategy of research* had not yet made its appearance at government level.

It was during the Second World War, and still more during the decade which followed it, that the idea of a *policy based on science* was forced upon the great powers: they initiated programmes of research aimed at the invention of new weapons. The considerable financial resources mobilized for these programmes have had the effect of putting the national scientific potential at the service of the geopolitical ambitions of the State, and of making the research institutions useful tools for reaching the targets set by governments.

The nations which had taken no part in the race for power or prestige, and those which exhausted themselves in an effort to keep up in this race, were understandably confused when they looked at the course taken by the scientific and technological explosion.

The fear of being out of the main current of this accelerating evolution, which everyone realized was of decisive importance for the future, therefore replaced the motivation of science for its own sake which raised no issues outside circles directly interested in the expansion of the budget for research. It is this fear of being left behind in technology—a fear, be it observed, as vague as that of being left behind in science—that has impelled the majority of the nations, especially those of Western Europe, to embark on nuclear and space programmes (national or multilateral, or both at the same time) without apparently understanding why or how they are acting, if one can judge by the state of confusion that these programmes are in after five or ten years.

The period of confusion has lasted over all the first part of the decade 1960-70, in the course of which numerous countries have provided themselves with an internal government structure for scientific and technological policy, and have created a variety of international organizations; they have not yet, however, formulated their objectives with such precision as to be in a position to draw practical conclusions concerning the line of conduct to be followed.

This is why a systematic analysis of the objectives of science policy has become necessary everywhere. It is being undertaken at this moment in numerous countries and several international forums. As a result of this, there is now gradually coming into focus a new vision even more ambitious than the 'policy based on science' of the power strategists; it is a *policy of economic and social development based on science.*

THE STATE OF AFFAIRS AT THE PRESENT DAY: SEVERAL SCIENCE POLICIES BEING CONDUCTED SIMULTANEOUSLY IN THE SAME COUNTRY

In some countries—especially in the small or non-industrial countries—the policy *for* science occupies the whole stage by itself. In these countries the government never pursues any objective other than that of procuring for its scientists, whatever the subject they devote themselves to, the material and moral support which they need.

The situation of the big powers is completely different. Policy based *on* science there reigns supreme, despite the efforts made to increase the 'fall-out' on industry of governmental programmes, and despite the fact that some authoritative voices are already advocating that the impetus towards progress afforded by government contracts for research, and government orders, should be spread over all the industries of the country. At the same time, these governments are paying more attention to subjects such as oceanography, the pollution of air and water, and technology of town-planning and management, etc. These are evident signs of *development based on science.*

Between these two extremes are found a number of countries where the three policies exist side by side and encroach on each other. Part of their decisions on science policy are inspired by the old idea of patronage and the duty of a ruler towards the scholars; concern for one's 'clients' in the Roman sense of the word, is still also found there, just as in the old days of patronage. Another part is bound up with the ambitions or the prestige of the nation. Lastly, a third part originates from an analysis of the social and economic situation of the country, and from the intention of making science a means of development. The greater part of the European nations, and a significant number of developing countries, are in the phase of indecision and change which characterizes the transition from one set of incentives to another. Nevertheless, one must be careful about pronouncing a negative judgement on this phase, since it probably forms a necessary stage of evolution.

It is likely that, without the malaise of the scientific community that can be ascribed to the inadequacy of the budgets and the structural organization of research, and without the geopolitical ambitions of the military general staffs, there would have been no analysis of the scientific potential of the nations, nor any comparisons among the countries from which today's science policy concepts have arisen. It is moreover useful, in the majority of cases, to start with a phase of limited science policy objectives, and to postpone expansion until the moment when the components of this policy have become sufficiently organized to serve national development effectively.

A policy *for* science and a policy based *on* science; does this distinction also contain a contradiction? True, it might be feared that, taken to the extreme, a policy for science would lead to the idea that the scientific potential of a nation is at the sole command of science; while, on the contrary, a policy based on science only considers science as an instrument of power. The one would lead to an ivory-tower type of scientific development, without reference to any other needs than those of the intellect. The other would lead to the subordination of the scientific community to political ends; it could even produce the situation where, whenever the activities of research scientists are not in accord with official policy, the importance of the role of these scientists is simply rejected out of hand.

The present situation shows, however, that the antithesis between these two concepts is more formal than real. For the resources put at the disposal of the scientists within the framework of a policy based on science are entirely disproportionate to those which they would ever have dared to hope for from a policy for science; and, besides, scientific developments have always been more rapid in the areas where concrete objectives are pursued.

Scientific research demands in return a tolerant intellectual climate and a freedom of inquiry and speech, without which fruitful investigation cannot exist. An industrialist or a government can certainly bend technological developments to their particular purposes. They cannot bend fundamental thought to them. If they try to do so, it fades away or ceases to exist altogether.

Whenever, in the history of science, authority has done violence to scientific freedom (as in the case of Galileo and, in this century, of biologists) its motives were ideological. No true planner would ever have acted in this manner, for if an administration occasionally finds itself obliged to make choices between several lines of research for which considerable expenditure is expected, and if in consequence it has to advance an estimation of the probability of certain scientific hypotheses, it is in its interest to keep all the avenues of success open to scientific endeavour. The ideologist is inflexible; a planner, on the other hand, has to be more prudent. The planner's task is to portion out the resources, not to pronounce judgement on scientific theories. He knows that he is bound to make mistakes, and that the wrong road costs very dear—not only in squandered resources, but in missed opportunities and time lost. His danger is that it may become evident by the success of other people that the business, country or community, whose scientific resources he administers, has blundered into researches leading to a dead end, and that he must immediately bring about a rapid reorientation of his programme. This will not be possible unless a team of scientists has been kept working on each of the hypotheses which have not yet been chosen, one of which has been shown elsewhere to be the best. He will therefore endeavour, by using caution, and by the grant of a minimum of resources, to keep all the branches of scientific inquiry alive, rather than deliberately sacrificing any. If the resources of the country are large, even the minimum can be sufficiently generously calculated. Such a policy is the rule in the fundamental sciences, especially the theoretical, where the material resources necessary are usually not very large.[1] Fundamental experimental science poses more delicate problems in allocation of resources, because particle accelerators, experimental reactors and orbiting observatories reach 'astronomical' costs, and here the rule of prudence is not so easily applicable.

It remains true that when the planner makes a choice (as he finally has to do), he does not pretend to be a judge of scientific truth, and makes no claim to substitute his own will for the liberty of the research workers. Today, far more than ever before in the history of financial aid to science, prudence demands of the planner that he should avoid neglecting even what appears to be the most improbable hypothesis, or the experimental research that seems farthest removed from the immediate aim; for he knows that what at this stage appears but a tiny path may subsequently prove to be the broad highway down which all will have to pass.

SCIENCE POLICY AS PART OF THE GENERAL
POLICY OF THE GOVERNMENT

Science policy is a part of general policy—the part which consists in improving the resources of science and promoting technological innovation to attain national goals. It has therefore very close links with other spheres of governmental action directed at the same objectives.

To explain these relations we shall here deal with the three principal spheres most closely related to science policy: education policy, economic policy and foreign policy.

Whether at first it is a matter of simply facing up to the accumulated deficiencies from which the scientific community suffers, or whether it is a matter of development based on science, the interface problems that occur

1. However, certain branches of fundamental theoretical science today demand a large amount of apparatus and therefore pose budgetary problems of a different order.

111

are not very different, and it is therefore important to set up from the start institutional machinery capable of dealing with them.

A symbiosis exists between research and higher education: the universities train research workers, and they carry out research. In practice this symbiosis is so strong that one cannot isolate higher education policy from science policy.

Between research and the economy the bonds are equally strong. At the outset it is merely a question of promoting the infrastructure of governmental scientific public services which the economy needs, including the institutions of agricultural research and co-operative industrial research. Later on it consists in formulating an integrated strategy for the stimulation of economic growth in which a *production policy* and a *policy of industrial organization* will determine a *research policy*.

National science policy also faces various problems of international relations: first, those connected with international or regional organizations having universal scientific or technological objectives, and next, those arising from bilateral co-operation agreements. For the 'super-powers', science is also at the centre of their geopolitical systems, because of the armaments race.

The place occupied by science policy in the general policy of nations may be modest, or it may be very large, but it will always be a central one. Indeed, science policy is situated at the point where three main lines of governmental action converge: education, economics and foreign affairs.

SCIENCE POLICY AND EDUCATIONAL POLICY

Scientific and technological progress entails a rapid transformation of industrial and professional organization at the national level and therefore induces a profound change in the needs of society for qualified personnel. Hence, its considerable influence on education and training, and on the management and utilization of their resources. Effective science policies obviously depend upon the number and quality of the human resources, which are among the essential factors of innovation. The educational system has to be developed and brought up to date in such a way that it can supply these resources.

Besides, education level and propensity to innovate are interdependent: an increase in manpower qualification throughout a society produces a situation conducive to innovation, which in turn produces the need for new qualifications.

The qualifications at the level of higher education are obviously the first ones that have to be dealt with, since they are directly involved in the tasks of research and organization, which are among the key factors of innovation. Furthermore, the progress of modern societies also develops large-scale needs for technical and administrative leadership.

The close relationship between the general level of a nation's education and its propensity to innovate is clearly demonstrated by a comparison between the educational level of the United States of America and that of the countries of Western Europe. University degrees are obtained by 7.6 per cent of the active population of the United States, compared with 2.8 per cent in the Common Market countries taken as a whole. In the United States, each individual spends on the average ten and a half years in the various levels of education. The average in most European countries is eight years.

This explains why the problems of mass education at the higher level are now clearly seen to be vital. These problems include forecasting the future needs of qualified manpower for production, research and administration. Qualitative problems, such as the content of education, may be of even greater importance, since the need to cope with and to control progress demands increasing knowledge and a fresh outlook on the part of individuals. A society which wishes to take up the challenge of change nowadays puts more emphasis on individual and social attitudes than on the stock of acquired knowledge. The need for well-balanced minds has never been so pressing as in this period of radical social change.

The success of every individual or collective enterprise depends on the ability to assess the changes which occur in the economic and social organization of society, and on the ability to interpret them no longer by reference to the conceptual frameworks of the past, but rather with the open-mindedness characterizing every genuine scientific endeavour.

A society wishing to exploit systematically its scientific and technological resources to attain its growth objectives must above all develop, in the education of its citizens, the faculties of fact-finding by observation, of adaptation to change and of rational decision-making. This amounts to saying that man must, more than ever before, learn to renew changing information and knowledge (because the positions won are under permanent challenge) and learn to make decisions (because the multiplicity of factors involved in decision-making is growing fast, which adds to the complexity of problems). These two conditions are essential on the one hand for fulfilling the tasks of planning and administration imposed by technological progress in a modern society, and on the other hand for ensuring a smooth mobility of human resources, which has become an important factor in the dynamics of economic activity.

In a country which remains at the stage of policy *for* science, the problem is mainly to achieve a balance between the supply of graduates with advanced degrees, and the R & D jobs to be created thanks to the increase of the national research expenditures. The expansion of university output must be rapid enough for the staff-supply to the research centres to remain smooth. The growth of these centres must absorb the flow of qualified research scientists which the university produces, to prevent it from being drained away to foreign countries.

In a policy of development based on science, the scope of the planning of higher education is larger; it is also linked to that of higher technological education, since its business is to provide the necessary highly qualified manpower needed in operative technology as well as in research. The supply of such manpower must emerge at the appropriate moment out of the educational 'pipe-line', and it must correspond in quality and number to the jobs which the national economy ought to have ready at the same time. Every

113

part of the educational system ought therefore to be adjusted not in the light of present needs, but in the light of the future needs of an economy whose configuration and organization will change while its future cadres are still at school.

The usual education span of the highly qualified cadres is ten years—i.e., the duration of two five-year plans—starting from the end of the first stage of secondary education, that is from the age of 15 to the age of 25. The education of qualified technicians lasts for five years, which corresponds to the length of one economic plan; namely, from the age of 15 to the age of 20. The planning of higher education therefore normally concerns the numbers of graduates with the necessary qualifications needed at the end of the second five-year plan after the current plan. Hence it necessarily relies on long-term options, to be selected in accordance with the overall national objectives. It is in the choice of these objectives that science policy plays its most essential role. Science policy-makers should thus extend their thinking to the plan for the development of higher education, which is of immediate concern to them.

Because of this, several countries have found it necessary to treat their policy for science, technology and higher education as a single whole and to name this whole for short 'science policy'.

SCIENCE POLICY AND ECONOMIC POLICY

In so far as research is geared towards development action, it logically expects the development policy to define its objectives. When research aims at the attainment of economic targets, one might similarly expect the research objectives to be defined by economic policy, which means that science policy would only come into the picture later, to supply its own particular contribution.

Those who plan and carry out science policy in many countries have accordingly asked the national development planning organizations to give them an outline of the national goals which they have to help attain. Answers have either not been given or only been delivered in very vague outline. This is why the inclusion of scientific planning in the over-all national development plan has so far only become effective as regards the *allocation of resources for research,* while the *choice of the general orientations of research* has often remained ambiguous.

The reason for this paradox is that scientific action requires a longer perspective than economic action, at least such economic action as is seen in five-year plans. We have already pointed out this fundamental distinction when speaking about planning for education. The comparison between scientific expenditure and economic investment also illustrates the difference in time-span: when a new factory, or road, or school, or block of houses, has been built, this represents 'plant' which is usually in use before the end of the plan in which the cost of its construction figures, so that the economic and social results of these investments form part of the final inventory of the achievements which were part of the objectives of the plan. But expenditure on research, like expenditure on higher education, produces its main results some time after the end of the plan, and must therefore fulfil needs corresponding to a final state of the system which the five-year plans do not define. The scientific plan therefore has to start with a more distant view of the future of the nation. In quantitative terms its outline will be less precise, but its qualitative content will need to be worked out with greater attention.

In fact, therefore, science policy outlines the long-term framework which those who are responsible for economic policy require. It anticipates the developmental changes which will take place in the economy in the course of succeeding economic plans. Since it affects key factors in the process of technological innovation, science policy necessarily extends the objectives of its operations farther into the future. Studies on the future evolution of the sciences, and technological forecasting, investigate a more remote future than that which is the object of economic forecasting.

One must take care not to regard science planning as a part of economic planning. The science plan cannot be identified with the economic plan, nor can it be completely integrated with it, except so far as concerns the volume of resources allotted to research as a whole, to higher education, and to the scientific public services. On the other hand, it is equally true that the two planning exercises cannot be realized separately. Indeed, a constant dialogue must be carried on between the organs of economic planning and those of science planning, so that the over-all development goals which the nation sets itself take into account all the factors, both qualitative and quantitative, of the growth process.

A CLEAR APPRECIATION OF THE POINT OF DEPARTURE AND THE STAGES IN VIEW

Such, then, would appear to be the general conditions of a development policy based on science.

For each nation, the premises of such a policy are to be found in a clear appreciation of that phase of economic development which the nation has reached. The scientific organization to be chosen at the national level is that which corresponds to the needs of the following phase of economic development, and which is capable of speeding up the country's entry into that phase. A small proportion of the human and financial resources can indeed be allotted to research which assists in completing and consolidating a previous phase or sector of economic development; or again, to research which fits in with some future phase in order to ensure the continuance of some special or traditional sector of the country's economy which may 'take off' when an appropriate situation arises. But the main scientific effort should undoubtedly be concerned with the problems of today and, above all, with those of the foreseeable future.

In this connexion one cannot over-emphasize the need for harmonization in time between the planning of research and the planning of higher education. Both are making ready for a future which is not concerned with the present situation and its immediate evolution. The first prepares the knowledge and techniques which will be used, and the second prepares the men who will apply them.

For its part, economic policy must use the intervening time to build the organization of production, goods and services for which the techniques and the men will have been prepared, and shape this organization into an appropriate size.

Fundamental agreement between science policy and economic policy is clearly essential. If the first reaches its objectives too quickly, the brains and the discoveries will depart for foreign lands. On the other hand, if the quantitative development of the economy is not accompanied at the proper time by a change in outlook, in output and in occupations, growth slows down or reaches its ceiling, and structural unemployment makes its appearance.

The authors of both the economic plan and the science plan must therefore share the same long-term vision. It is the group in charge of science planning and planning for higher education which is most in need of this vision, since these derive their immediate targets from that vision. Since this group also has at its command instruments for assessing qualitative change, it is therefore logical that it should be charged with the task of studying and putting forward the nation's long-term development goals, in conjunction with the group responsible for economic planning.

SCIENCE POLICY AND FOREIGN POLICY

We shall pass rapidly over the links between science and military strategy, because this problem only affects the 'super-powers' and a small number of medium powers who have maintained ambitions of the same kind.

However, these countries share with all the others the vast number of links between civilian research and foreign policy. Four examples of this are briefly dealt with below.

Exchanges of professors and students, and mutual assistance in the cultural and technical fields

This kind of co-operation has a very long history. It underlines the spontaneous tendency towards unity in science throughout the world—a tendency stemming not only from the aspirations of the community of scientists themselves, but also from the universality of the scientific approach to problems. This unity is of long standing. In Europe it dates back to the high Middle Ages, a period when discreet links had been forged with Arabic science across the Mediterranean. It spread to the other continents, under the pressure of ideas which had their source in the scientific and technological revolution. Since that time no ideological or political quarrel has succeeded in really breaking those links, even when communication between individuals has been made difficult.

Many countries, far from hampering this unity, promote it by giving aid to visiting professors, to students and probationers, to the organizers of meetings and conferences, etc. This is the most elementary form of foreign policy in scientific matters, and even today it remains of considerable importance.

Co-operation in the fields of research and scientific public services

Several countries may pool their resources in order to carry out specific research together, or to set up a scientific public service. Several nations may associate together to adopt this expedient, when the resources of each are insufficient to supply the machinery or teams of research-workers which are necessary.

Chapter 8 deals in detail with problems of scientific co-operation.

Technological and economic integration

At the beginning of the second phase of industrialization, problems of scale confront countries with a severity that has not been experienced in the preceding phase. They are forced to group themselves in partnerships, with the aim of creating together an economic, technological and political system of the necessary size, and to form a close association which in turn determines a common destiny. This phase can be guided or speeded up by the shortcomings of the previous co-operative approach, if the current of ideas prevailing in the Western Europe of today can be taken as a precedent.

The transfer of technology

The fourth in our list of problems is that of the horizontal transfer of technology, the possibilities open to direct investment by foreign enterprises, and the protection of inventions. Here a nation is confronted with a major choice in establishing its economic relations with the outside world.

Theoretically there are several ways open for choice.

It can import technology in the form of 'turn-key' factories. It is usually stipulated in such contracts that the contractor must start the factory up and train its labour force, thereby including an educational element in the contractual arrangement. These contracts do not involve definite subordination of the new factory to the foreign owner of the technology to be transferred. This formula therefore assumes that the country will take into its own hands the technological improvement beyond the stage reached by the provider of the equipment. It assumes that there will be a prolonged effort of technological research, either by the importing country or by the newly-created industry, once production has been started.

It is also possible to import technology by direct investment, which means the building of a factory owned by a foreign company, or by some company under its permanent control. In this case, the parent organization continues to nourish its offspring with new technology, as and when this is produced in its laboratories, which are usually to be found in the country of origin. The host country is freed from any anxiety on this score. Nevertheless its dependence will remain permanent.

It is equally possible to buy a patent or take out a licence for processes or products, if one wishes to start an autonomous industry.

The last possibility is to refuse legal protection, inside a country's boundaries, to foreign inventions; so permitting the use of such inventions for internal needs without having to bargain for licences or to pay dues. Naturally this policy also has its limitations, since the production thus achieved cannot be exploited on the markets of the countries which adhere to the Paris Convention on the protection of industrial property (which includes all the industrialized countries in the world).

The choice of one or other of these policies clearly depends both on the will for technological autonomy of the country concerned, and on the practical possibilities open to it either of winning such autonomy or of preserving it.

It is worth noting here that the agreements for the construction of 'turn-key' factories are usually only offered for basic and traditional industries. In science-based industries, the providers of the technology have a very clear preference for the formula of direct investment, and they often refuse to negotiate licences unless every other possibility is closed to them, or unless an exchange of technology or 'know-how' is offered. When they are negotiating with a local firm which does not possess a scientific and technical potential capable of offering a fair exchange of patents and 'know-how', these providers usually insist that the bulk of the shares should be handed over to them.

On this delicate subject of the conditions for the horizontal transfer of technology and direct foreign investment, science policy, foreign policy and industrial policy coincide, and they should be closely integrated.

CONCLUSION

Science policy is at the cross-roads of economic policy, social policy and foreign policy and is therefore the responsibility of the whole government and—by force of circumstances—the responsibility of the head of government.

Because of the inadequacy of organization and finance which hampers scientific workers in the majority of countries, science policy has gradually widened its scope by reason of the very nature of the questions which it has posed for governments. In the countries which have chosen a policy of development based on science, this policy fulfils a function of long-term orientation of the nation's over-all development. Meanwhile science policy has never ceased to be concerned with the career opportunities and working conditions of the research workers, and with the promotion of research and development of higher education.

In the course of this evolution, the demands of policy-making—a staff function which is a distinctive feature of management at the highest level—have obliged science policy-makers to lean away from the tasks of execution and the responsibilities of routine management, in order to concentrate on the choice of objectives and the allocation of resources.

7 The functions of a science policy

INTRODUCTION

It may be useful to recall at this stage that science policy is not science itself. It serves science, and from science in return it expects services for the good of the community. Science policy must not be confused with the performance of scientific work. With the progress of society as a whole in mind, science policy sets itself the objective of offering to scientific activities the optimum conditions for their development.

Upon closer examination, such a policy may be said to comprise four functions: planning, co-ordination, promotion and execution.

The aim of *planning* is to define the essential objectives, to decide their order of importance, and to determine the resources that have to be mobilized to achieve them.

Planning is essentially a strategy of options, and therefore depends upon correct information. Planning also implies the building of a consensus among the representatives of the different interests involved; finally, planning demands efficient decision-making machinery.

Co-ordination between government departments is an integral part of science planning and of putting programmes into operation. It is first practised for the benefit of the programmes themselves, so that all the objectives selected in the course of planning are taken into account, both at the time of initial formulation of the programmes and on the occasion of their periodical revision and adjustment. It is next practised when the programmes are carried out, in order to obtain proper operational harmony in the achievement of the objectives selected.

Being a strategy of consistency, co-ordination has a vital need of liaison techniques both within the government, and between government and intermediary or semi-governmental bodies.

Promotion creates the necessary conditions for the realization of the objectives. Action is initiated by granting the necessary resources; and evaluation ensures that the use of these resources produces the expected results. Such evaluation helps in the periodical readjustment of the programmes.

Promotion is a strategy of action; its essential function is to put into operation, within the framework of the selected programmes, various techniques of financing and of evaluation of the results obtained.

Execution aims at the practical realization of programme objectives. It combines human and material resources, which are part of the scientific potential of the country, for the creation of new knowledge, new products, and the new machinery and equipment which will make it possible to achieve the selected objectives.

Here we come to the level of tactics, where the resources are put to use. At this level the instruments are management techniques, the goal is optimization of resources, and the agent is the research director.

FIRST FUNCTION: PLANNING

Being a strategy of options, the aim of planning is to define objectives, to decide their order of importance and to determine the means which have to be mobilized to achieve them.

CHOICE OF OBJECTIVES

The general approach

Every choice of objectives springs from the will to act, that is, to change something in the present situation. But action takes time, and the world changes rapidly. It is therefore misleading to determine the objectives of research by relating them to the present-day situation. They must rather be related to what the situation of the country would have been at the envisaged time of completion of the action planned—had this action *not* been undertaken. Just as the hunter aims at the place where the animal will be when the shot reaches it, and not at the place where it is at the moment of firing, so the planner bases his thinking on a future situation.

The first stage of this process is therefore to forecast the probable course of events. To do this it is necessary for the planner to know the present situation, the meaning and intensity of the changes that have happened in the recent past, the internal driving force of the present system (that is, its dynamics at the moment), and the probable incidence of external factors. From this he forms a picture of the future of the system as he foresees it. We shall return later to techniques of forecasting. At the present stage what really matters is to admit that such a picture can be constructed and that it must be submitted to *critical examination* to begin with.

Thereafter the planner can either accept this picture, or change it for a better one. This picture then represents the *development blueprint* which he puts forward for governmental action, due account being taken, in drawing up the blueprint, of the government's general directives. The blueprint hence depicts what the *final state of the system* will be, once the proposed national development plan has been effectively implemented.

This *proposed national development plan* will be considered to be theoretically possible if, between the initial and the final states of the system, a continuous series of intermediate stages can be defined in such a way that the transition from each to the next does not use up more of the resources (human, material and financial) than the society has at its command during the phase in question. The analogy between a plan of action and a physical model is obvious here; in the former, resources play the part which energy and mass play in the second.

Not all the paths which thermodynamics postulates as possible for a system to transit from one state to another will be put into practice by the engineer; in the same way, not all the plans that are theoretically possible are realizable. The mobilization of the resources of a people encounters practical difficulties and losses of efficiency, as in the construction and operation of a machine. It is therefore the politician's task to fit the pace to the level of the resources that can be effectively mobilized, and in particular to adapt it to the tempo and the sequence of the reforms which can in fact be pushed through. He may even possibly feel obliged to downgrade, so far as his sense of prudence makes this appear necessary, the objectives of the proposed national development plans, in order to ensure that development as a whole shall not be endangered by accidents *en route*.

The choice of the system of values, from which the *development blueprint* springs, as well as the assessment of the difficulties and dangers of the action, are political decisions; the technocrats or managers can pave the way for such decisions, but they cannot take responsibility for them. They can no more dispense with the politicians than the politicians can dispense with them. However, collaboration between the two presupposes that each will have acquired a thorough understanding of the other's function.

Now that the conditions under which the choice of development goals takes place have been thus specified, let us examine factually the way in which the general objectives of a development plan based on science are chosen in practice; we will begin with the economic objectives.

The phase of development which a country has reached determines the nature of industrial production

An economic plan, in its usual scope of four or five years, aims to increase the production of all the goods and services which are already being produced at the start of the plan, and in addition to produce a certain number of new commodities, and services, whose manufacture has been decided on. But in practice these already exist and are known, either in foreign countries or in the laboratories of the country itself.

On the other hand, a long-term plan of development aims at products and technologies which do not yet exist in the country, and, sometimes, do not even exist elsewhere. As for the scientific plan, its aim is to evolve these products and technologies.

It will therefore not be possible to produce the scientific plan until a rough outline has been made of a long-term plan of production, or at least until there has been some preliminary thinking about such a plan.

This thinking will give an indication of the main axis of the national effort of investment, production and export in the next twenty or thirty years. The spheres of technology in which it is essential to obtain either new break-throughs or a level of excellence and competitive strength will become clear from this. From the technological priorities it will then even be possible to draw conclusions about the orientation of fundamental research.

As a start, therefore, the following question will be asked: of what kind of goods and services must production be forecast?

Except for the unique case of the United States of America, a first rough estimate can be made by studying the analogy with the past situation of nations which have gone through the same phase of development earlier. Yet no one will expect to see the nations of Asia, Africa or Latin America pass through in their turn all the stages of economic and social development which Europe and the United States of America encountered, and even less all the historical stages that technology and science have gone through in those countries. The developing countries which have entered the race later take a series of short cuts, which offer them moreover the prospect of catching up with the advanced countries. The short cut is easy to find so far as the body of knowledge is concerned; it is also easy in the choice of opera-tive technology.

But so far as the choice of the products themselves is concerned, the stages of economic development limit the choice more severely, because the demand of the consumers changes in quality in proportion as their purchasing power increases (Engel's law), their purchasing power following the same curve of increase in the graph as GNP per inhabitant. This is why a sequence, food—clothing—household equipment—private cars—articles of leisure, is neces-sarily repeated in the growth of the consumption pattern of all peoples. In fact it corresponds to a natural sequence of the needs of any population, in order of decreasing urgency. This sequence can be modified in certain instances, either by climatic conditions or local cultures, or by the example of more advanced countries (here one is thinking particularly of the place occupied by wireless and television) but it cannot be altered very radically.

Since the choice of consumer goods to be produced for the needs of the nation is divided up into necessary stages, the choice of intermediate goods and capital goods is also divided up in the same way except in the case of raw material substitution. For example, the needs for domestic heating, cook-ing utensils and clothing will develop in their proper place in the sequence, but some newly industrialized countries may use oil, aluminium and man-made fibres when similar needs in the past were met by coal, cast-iron and wool respectively in the early developed countries.

With this reservation, the national needs corresponding to the current stage of development will largely control the nature of the products and technolo-gies. To the limiting influence of this factor must be added that of internatio-nal trade, each country developing the products which give it the best terms of exchange, or products aimed at import substitution.

And so, on the basis of comparative costs on the one side and the priorities of the national balance of accounts and payments on the other, an increas-ingly sharp division of labour emerges between countries in their first phase of industrialization and those in their second phase. The former come to operate more and more in the markets for intermediate goods, textiles and other products of everyday consumption, the latter to specialize in capital goods and in products of advanced technology. However, as a country becomes more industrialized and better educated, it diversifies its production and in the end supplants countries which preceded it in some markets of machinery and chemicals, where those countries thought themselves to be firmly established. By this progression from the simple to the complex, the less advanced countries can make up their lost ground. During the next three or four decades, the *law of comparative costs* will thus bring about a new division of labour between the nations, as a result of the differences in their level of development. The traditional format of the nineteenth century and first half of the twentieth (the exchange of agricultural products and primary materials with the whole gamut of industrial products) tends to give place to a new arrangement: the exchange of industrial products of the first generation with the machines, appliances and products of advanced technology.

The phase of development governs the orientation of technology and the priorities of the scientific plan

It could be put forward as a general guideline that the problems towards which the main technological efforts of a country are directed ought to correspond to the types of production proper to the next stage of industrialization. In this way, the next stage can be efficiently prepared for and can be completed in a fairly autonomous way, since it will only require a limited amount of foreign assistance.

Although fundamental research ought not in principle to be limited by the current constraints of economic development, except perhaps when the cost of scientific equipment is very high (particle accelerators, space satellites), it is clearly desirable that a country's potential in pure science should be closely connected to the technological orientations which it has chosen. Daily contacts between the scientists engaged in basic and applied research have to be maintained, in order to ensure an appropriate cross-fertilization of ideas. In any case, pure research only absorbs a small fraction of the scientific resources of a nation, and there is therefore no need to plan how it is to be used, except when the need arises to mobilize all the energies of the country to achieve a particular break-through. Apart from nuclear weapons, the history of science contains few outstanding examples of this. Most governments therefore refrain from planning pure fundamental research; moreover, this would rouse justified objections on the part of the scientific community.

On the contrary, the choice of technological orientations as a function of the products that correspond with the next phase of a country's economic development constitutes a characteristic feature of governmental policies aiming at economic autonomy.

In order to be able to make these choices, a country must adopt a clear-cut *development strategy*.

Can a development strategy dispense with qualitative progress?

In the first chapter we were led to distinguish five different development processes: (a) extensive growth; (b) accumulation of fixed capital; (c) improvement of structures and organization; (d) horizontal transfer of technology; (e) original technological innovation.

The first two processes comprise a quantitative increase of the factors involved in production. Here, the national effort involved makes no call on science as an instrument for action.

The last three processes, however, demand a qualitative change in the methods of production and their social context. Here, the application of science and technology to development operates in ways described below.

The improvement of national structures and organization is a method of growth which does not involve technological change, therefore it does not require, of operatives, new acquisitions of technical knowledge. It is theoretically obtainable without raising levels of education among workers, and without technological research. This process of growth nevertheless does require scientific methods of decision-making and management.

The horizontal transfer of technology (the fourth process) requires assimilation of foreign knowledge and 'know-how'. This usually implies that workers and cadres in the recipient country should reach higher levels of education; in certain cases, considerable efforts towards improving the cultural equipment of the people are required. For a technological revolution to make profound penetration, a systematic effort to educate the masses must generally be developed over a long period.[1]

Finally, original technological innovation (the fifth process) demands a national research effort and the creation of a *potential for technological invention*. Here, the demands for qualitative progress are far greater than in the case of horizontal transfer of technology.

One might be tempted to conclude, from this enumeration of difficulties, that priority should be given to the simpler processes of economic growth until the possibilities that they offer are exhausted. In this way, each country would pass through successive stages of development—quantitative at first, and later qualitative. Under such a hypothesis, however, only the United States would need large-scale original technological innovation supported by systematic research and experimental development as a method of growth; countries that are already industrialized, but whose *per capita* production remains significantly lower than that of the United States (like the countries of Western Europe), would rely primarily on the transfer of American technology; and the developing countries would concentrate their energy on the three first processes of growth, which demand no special efforts on the part of either education or science. The analysis which follows will show why this seemingly logical sequence is not followed in reality, and why there is clearly no good reason to advocate it.

1. It has been pointed out that the 'second industrial revolution' actually involves, in its effect on secondary and university education, an expansion in the school building programme which is comparable to that which the first industrial revolution required in primary education.

The demands of the argument make it necessary for us to deal first with situations prevailing in the second phase of industrialization.

Development objectives during the second phase of industrialization (Europe, U.S.S.R., Japan)

Study of the present economic development of Europe shows the five processes of development, in combination as *industrial entrepreneurship.*

An example will illustrate the integrated complexity of its nature. A chemical firm is passing from the stage of pilot-plant experiments to the construction of a large-capacity factory for the production of a new plastic which is meant to replace several traditional products.

This usually creates new employment, over and above that which has been lost in the superseded industries. The first process of growth is therefore present since the working population increases.

The fixed capital deployed in proportion to each job in the new factory is much higher than it was in the earlier factories. The added value per worker is also higher and its growth is faster than the increased consumption of fixed capital. The second process of growth is thus also included.

The new factory, operating at a high productivity level, has replaced two or three smaller factories which used to manufacture traditional products in a less systematic kind of organization. It may also have made it possible to close down some coal mines, farms and retail shops which had lost their profit-earning capacity. The industrial structure of the nation and the organization of production have thus been changed. This is the third process of growth.

The product is new in the country, and the operative technology is also new. A licence has been bought from abroad or obtained by an exchange. So there is innovation by horizontal transfer of technology. This is the fourth process of growth.

Finally, in the course of the pilot-plant experiments, the company has improved on the original patents, for which it had taken out the licence. It has thus shared in original technological innovation, which is the fifth process of growth.

A single act of investment, the building of a new chemical factory, has therefore put the five processes of growth into operation simultaneously. Since they are present at the same time, we now call them *factors of development* rather than processes of development.

No doubt there can also be found, in industrialized countries, acts of investment where one of these factors is absent. This is notably the case in the rationalization of old industries, which are made competitive again by labour-saving investments. The first factor is therefore missing. This is also the case when foreign firms establish branches in a country for the purposes of production only. The fifth factor is clearly absent here, since research and experimental development and a host of other activities which constitute the process of technological innovation are carried out abroad in the parent firm, and will remain there. However the second, third and fourth factors of development are closely combined in these two kinds of industrial achievement.

It will be concluded from this analysis that the industrial decisions, conducive to progress in industrialized countries, usually include the combination of at least three of the five above-mentioned factors of development. It is impossible, therefore, to dissect the successive stages in the development of these countries. Industrial progress is indeed an *integrated process*.

It will be noted also that the weak points in the economy of industrialized countries are precisely those industrial operations which are incomplete, that is, those which do not create new employment and which do not include original technological innovation.

Let us first take the case of the operations of rationalization in the traditional industrial sectors. The increase of capital outlay per job is accompanied by the discharge of a proportion of the labour which is not re-employed. Reorganization is also incomplete: unprofitable production units are abolished, and profitable units are enlarged and modernized, but no new unit is created with a new orientation. For this reason, even when an increase in the gross regional product is gained (as is usually the case) the process is incomplete, and does nothing to prevent an industrial regression in the region concerned.

The establishment by a foreign enterprise of a branch for the sole purpose of production may introduce into the region new employment in a new industrial sector, and a positive change in the region's industrial structure. However, the unit of production thus established is neither an autonomous economic enterprise, nor a creative member of a multinational business organization. It remains a passive and subordinate instrument of the foreign enterprise which established it. So it will not become a source of original technological innovation and industrial entrepreneurship for the country or the region which accepted its implantation. It is unable, in its turn, to set in motion again the process of development.

The depressed areas of Western Europe, most of which are the old industrial basins established in the nineteenth century on the base of the coal industry, show a predominance of these two types of incomplete industrial operation: local investments in rationalization of traditional industries on the one hand, and on the other investments by foreign enterprise based solely on the horizontal transfer of technology.

The restoration of a complete and healthy process of industrial entrepreneurship with all the five factors of development present appears to be an essential condition of economic vitality in those industrial countries. In such a process, research and experimental development play a decisive role since they are the source of the innovation process.

*Development strategies in the course of the first phase
of industrialization*

For the sake of analysis we shall first consider the successive use of the five processes of development by way of hypothesis and as a theoretical pattern of development.

The characteristic features of the economy of countries not yet industrialized are usually the presence of an unemployed labour force and a prevailing
lack of fixed capital. It would therefore appear to be logical to begin by
employing this reserve (first process) and by a generalized investment policy
(second process).

Such a succession of stages was possible in the past, in the periods when
technological progress was slow and the contacts between civilizations which
had reached different levels of technical knowledge were rare. In those times,
the assimilation of foreign techniques and original technological innovation
both presented considerable difficulties and hazards. Supposing that there
could have been something in ancient societies which took the place of what
we now call a policy of economic and social development, priority would
perhaps have been given to the quantitative and organizational factors of
development in preference to the qualitative factors.[1]

At all events the reasons for such a priority could have been the difficulty
and the hypothetical and hazardous character of qualitative progress. It is
likely that the authorities would have considered the structural reform of
production easier than technical progress. But, in our epoch, horizontal
transfer of technology has become easy, whereas reforming the structures of
traditional production, and providing employment for the underemployed
majority—even if these steps are carried out at a very low level of capital
investment—always present considerable difficulty. Industrial innovation
based on imported technology is by comparison a relatively easy and swift
way. This is why one sees modern industries, with a high level of capital
investment per job, springing up in the developing countries. These industries
use only a part, and often a small one, of the national labour force, while the
mass of the population remains working within traditional social and industrial structures with but antiquated equipment. There are large differences in
the productivity—and often in the pay—of the two parts of the working
population.

Variations like these in technological progress may be explained by the
decision—whether by private investors or the public authorities—to apply
simultaneously at least four of the five factors of development of production
(recruitment, capital investment, organization, horizontal transfer of technology) to a limited number of workmen rather than to apply the four methods
successively to the whole working population. The problem is whether such
a decision ensures the optimum use of the available resources: in other words,
whether the total growth of the domestic product would not be greater if the
same efforts had been devoted to the recruitment, training and equipping, at
a lower level, of a larger number of workers, using the framework of traditional techniques, but a reformed industrial structure. The problem assumed
very large dimensions in mainland China at the end of the decade 1950-60.

1. The example provided by Tsar Peter the Great just before the first industrial
revolution suggests, however, that the qualitative method was sometimes preferred
even in that period.

This problem is vital to our present purpose, because if the priority usually given to technological progress were called in question, so too would be the general applicability of a development policy based on science.

The answer is far from clear. No developing country appears to have avoided the dichotomy between a modern sector and a traditional sector, both in agriculture and in industry. Such a dichotomy is met in every example of an applied development policy, and mainland China, although it has the merit of having faced up to the problem, is no exception to the rule.

The central issue appears to be not so much the need for a modern sector, as the proportion of the total investment it deserves in comparison with the share of investment which ought to be directed towards equipment, structural reform and the horizontal transfer of technology in the traditional sector. The proportion varies according to the development policies adopted. The search for an optimum ought certainly to take account of the *driving force effect* which the modern sector has on the traditional one and, to give the opposite side of the picture, the *effects of disruption and isolation* which can be the results of too great a disparity between the two sectors.

Our provisional conclusion will be, therefore, that the successive application of the processes of growth is not to be regarded as a general pattern, even when there is a large number of underemployed—the first four processes will always be found together in at least one part of the economy: its modern sector. In developing the traditional sector one can perhaps, in a first phase, allow larger scope for the quantitative factors of development (the two first processes) and for structural reform (the third process). But it is difficult to admit that horizontal transfer of technology could be dispensed with in such a phase.[1] On the contrary, the dissemination and use of scientific knowledge in rural areas will always be an important development factor, even if the transition is not made at once to the use of modern equipment and sophisticated manufacturing. Full transition would in fact demand an outlay of fixed capital per inhabitant and a generally high level of education which could not possibly be attainable at short notice.

There remains the fifth factor: original technological innovation based on research and experimental development. Is this necessary for countries which are not yet industrialized? Would it not be better for these countries to do without it and build their modern sector exclusively on technology transferred horizontally, thus saving the nation the expense of research?

The answer to these questions will finally settle the place of technological research in the patterns of development of the non-industrialized countries.

The majority of the experts are of the opinion that participation in original technological innovation is a necessary component of every national development plan, without exception. They base this answer not only on the criteria of technological autonomy, national dignity or prestige, although these considerations are given some weight in reaching the decision. Five technical justifications for such an opinion are usually put forward:

---

1. The case of the artisanal iron industry in the rural areas of mainland China illustrates the limits of an effort based exclusively on the first two processes.

1. The horizontal transfer of technology, by itself, cannot give the correct answer either to the agricultural problems, which are peculiar to the local climate and soil, or to the mining problems which are peculiar to the local geological setting. There is always a need for a national effort of research and experimental development in order to facilitate the vertical transfer of technology in these fields.

2. Just as the transmission of a message by wireless needs a receiving station tuned into the same wavelength as the transmitting station, so the horizontal transfer of technology is successful only if there are present in the receiving country teams of research workers and engineers who will act as the spearhead for the application of science and technology to development; it is clear that these teams will not reach or maintain high standards, unless they themselves share in the process of original technological innovation.

3. The horizontal transfer of technology obtains better conditions by barter than by purchase: only the country where industry can in exchange produce results from its own laboratories will in fact be able to get the best foreign licences and have access to the most advanced 'know-how' of the outside world.

4. Every stage of development must be a preparation for the following stages. Thus in the course of its first phase of industrialization, a country ought to start building a firm base from which to set out on its second phase. It ought in fact to possess, at the end of its first phase, a thriving potential for original technological innovation, at least in some well-chosen directions.

5. The *scientific public services,* which play an essential role during the first phase of industrialization, cannot reach and maintain the necessary standard of quality without a considerable effort of research, nor can they fulfil their role of education in the nation. As we have already emphasized, scientific practice will be understood and assimilated only if it becomes an integral part of the life of the nation from beginning to end, from fundamental research to decision-making in industrial entrepreneurship.

The premises which have just been laid down make it possible to suggest four key objectives for a development strategy which is suitable for the first phase of industrialization:

1. To make a speedy increase in the productivity of agriculture and mining in order to ensure, by means of the high degree of viability attained in this sector, an expanding internal market; this will absorb the production of the growing industrial sector, and will release for industry a major part of the labour force which agriculture was employing in conditions of low productivity.

2. To speed up the horizontal transfer of technology with the aim of introducing industries of the first generation and of creating for that purpose, inside the country, some potential for original technological innovation, in order to build a modern sector of industry; care must be taken that this modern sector should be grafted on to the traditional economy, so that it can act as a spur to that economy by its purchases and sales.

3. To complete the infrastructure of the scientific public services which are needed to speed up industrialization of the country.
4. To begin to create a national potential in certain spheres of advanced technology in such a way as to lay the foundations for strongpoints in the future; by means of these it will be possible, at the end of the first phase of industrialization, or at the beginning of the second, to start setting up autonomous science-based industries. The objectives of a scientific plan can be easily deduced from such a development strategy.

The development strategy of the American economy

At the stage in the argument at which we have arrived, we can conclude without hesitation that no development strategy exists in which original innovation, that is research and experimental development, does not play an active role, although its points of impact may be very different. It is only necessary to examine in detail the case of the economy of the United States of America. It is clear that technological innovation is the chief activator of its growth.

However, horizontal transfer of technology also plays what is, at first sight, an unexpected role in this economy. If the exchange of scientific and technological knowledge could be measured in the same way as the exchange of goods, it would certainly become apparent that the United States is a large exporter of this 'commodity'; but what would be more surprising is the fact that it is also the chief importer of scientific knowledge and 'know-how'.

This paradox arises from the fact that the progress of different countries, like the progress of their various industries, is not made in the same spheres and on all the scientific and technological fronts at the same time. That industry or country which is the furthest advanced over-all, is by no means first in everything. It is true that the world's research potential is very largely concentrated within the group of industrialized countries; and it is also true that a great part of the scientific resources of that group is concentrated in the most highly developed country, namely, the United States. Nevertheless, scientific discovery and original technological innovation have remained or become part of the life of every country in the world to a greater or lesser extent. The sum total of the world's discoveries and technological innovation represents a potential for progress which is available for all.

It is, however, clear that the United States is in a better position than any other country to use this world potential to the best advantage in order to supplement the developmental drive which it obtains from its own national potential for research. Its superiority is found on three planes: (a) management of technological innovation, (b) characteristic features of the market, (c) support which industry receives from the Federal Government.

The management of technological innovation. United States industries, whose managements pay very careful attention to technological change, are organized in such a way as to keep themselves properly informed about everything of importance to them which happens in the universities, the research centres and the industrial enterprises of the whole world. They offer favourable conditions to foreign inventors who bring them the fruits of their efforts.

Further, it is a long-standing characteristic of the American people always to take particular care in the practical application of new knowledge, and in the commercial exploitation of the product which is the result of this application. This trait of the national character is given splendid opportunities by the capacity for organization and financing of new projects which the large American firms have developed in the process of technological innovation.

But the qualities of efficiency, energy and strength which characterize the management of this process in the United States of America must not make us forget either the stimulating effect of the most important market in the world or that of the most powerful government.

The characteristic features of the market. The American market presents several favourable characteristics which the European Common Market does not possess, or, at least, not yet. First its very size, expressed in billions of dollars of total annual expenditure, is about twice as big. Secondly its integration: it is not divided up into compartments by regulations or technical rules applicable to small areas within its borders. Owing to the existence of the Federal Government, the limits of the common market coincide with those of the power which issues regulations. New products can be marketed immediately over the whole extent of the American market, and their manufacture assumes from the outset a suitable scale. Thirdly the market is composed of consumers who possess a considerably larger average income, and of industries which are expected to pay wages graded in the same proportion. Such a market is more open to new equipment and appliances. To be more accurate, it is open to a given type of equipment or product several years ahead of the other markets of the world.

United States industry does in fact give earlier proof of the need for sophisticated labour-saving devices, and the individual North American citizen reaches at an earlier age the level of income which allows him to buy domestic appliances of a new kind. For this reason the effective market for durable consumer goods and advanced equipment is open in the United States ten to fifteen years before it opens up in Europe. There is a similar time-lag between the date of the opening of the effective market in Europe and the developing countries.

An inventor who is not an American and wishes to launch a product of advanced technology on the market of his own country therefore finds the immaturity of his home market an obstacle. There is a big temptation for him to offer his project in the United States, in order that it be developed and marketed with the maximum chance of success. Since firms there are geared to accept this kind of initiative, and even to attract it, a large part of the creative genius of Europe and the Third World finds its way to the most highly developed economy; the results may return later to the country of origin in the form of licences or direct investment by United States firms.

The support of the Federal Government. The purchasing power of the United States Government, and the vast size of its programmes in the fields of defence, atomic energy, space and oceans create openings in the country for the most advanced products and techniques long before these are ready to be offered on the commercial market.

The combination of these three conditions gives the United States enormous attractive power for industrial invention and scientific research. The power of this attraction lies at the origin of the brain drain into this country. If it remained unchallenged it could, in the long term, weaken and perhaps wipe out the centres of original technological innovation which are still active in Europe, in Japan and in the developing countries. The remedy is clearly to be found in an active policy of technological innovation on the part of the countries which are threatened by this phenomenon.

The size of the nation, and the development strategy

We have just shown that development forms an integrated process in which original technological innovation and therefore research are everywhere important factors.

In countries where the horizontal transfer of technology and the reorganization of the industrial structures to deal with larger markets appear, at first sight, to be expected to play a large part, the most healthy industrial initiatives are those which combine the five factors of development and therefore include original technological innovation. They are also the only ones which safeguard national independence.

In developing countries, as much as in others, participation in research and original technological innovation is indispensable not only to meet certain specific national needs, but more especially to open up and maintain the lines of communication with more advanced countries through which the horizontal transfer of technology occurs.

Although a development strategy calls in any case upon science, it will not necessarily require an identical type of research or an identical scientific organization. The size of a nation and the phase of development which a country has reached will affect the problem in differing ways.

In countries which have entered the second phase of industrialization, the chief objective will be the expansion and competitiveness of the science-based industries. The quality and size of industrial structures, their potential for technological innovation, their dynamism, the size of markets, and the support which government contracts for research and equipment can give, will be essential factors in their success. New constraints will then appear at the level of government itself, because the small or medium-sized nations, which have reached this stage of development, will necessarily have to regroup so as to form together an integrated market—and also a federal or confederal government, or at least a community that foreshadows such a government. If they fail to reach the size-thresholds necessary for large-scale technology, thresholds which have to be crossed at the same time both in the industrial sector and in the government sector,[1] the technological objectives of the

1. In the meaning given to this expression by the macro-economists. They distinguish three sectors in the whole of society: the industrial sector, the household

development strategy which the situation of these nations requires them to pursue would be beyond their power, and their economic progress would thus be inhibited. If this inhibition is not overcome, the nation will certainly be swallowed up in a more powerful economic system;[1] this can happen without its losing its formal independence, but not without its handing over its autonomy of decision and its freedom of action in the world. Acute symptoms of this can be seen in Western Europe, where nations of between 10 and 50 million inhabitants present a framework whose scale is inadequate for the second phase of industrialization. The formation of a common market has been the first response to this situation. The setting up of a technological community based on consortia of industries and a common policy of contracts for research and experimental development or government orders is the continuation, without which these economies would be inhibited in their growth and would before long be captured in the orbit of the strongest industrial system.

In countries which are carrying out their first industrialization, the problems of the size of a nation[2] are not so critical. The example of the Grand Duchy of Luxembourg shows in fact that no size-threshold was encountered during the development phase (now left well behind) of the economy of this country—an economy characterized by heavy industry. Even nations with a very small population (the example of Mauretania springs to mind) can therefore create competitive units of production in the basic industries of the first generation, if they have access to the world market.

At the time when the second phase of industrialization comes in sight, the nations of small or medium size will experience, in their turn, the problems of international integration which Europe is experiencing at the present time, problems which today are vital for the Grand Duchy of Luxembourg, Belgium, the Netherlands, etc.

The size-threshold which is necessary for the autonomous development of nations during the second phase appears today to be more than 100 million inhabitants. But thresholds vary rapidly, and before long this could be more than 200 millions. In the long-term perspectives of economic development, this is a geopolitical factor which must not be overlooked.

---

sector and the government sector. The latter includes all the powers of local, regional, national or federal governments and all the government services (with the exception of those whose output is commercialized, which come under the heading of public enterprises, and therefore belong to the industrial sector).

1. Such systems (in the sense adopted in modern systems analysis) can take shape through the intervention of multi-national enterprises. These enterprises, although they depend financially and politically on the most powerful government in the system, develop their production and trade in all the countries under the influence of the latter. The obstacles caused by the size of a nation are thus effectively overcome in these countries, which are henceforth united in a working industrial system, adapted to the exigencies of the second phase of industrialization. They clearly occupy a position on the periphery of this system.
2. It is the size of population which is important and not the extent of the territory.

Non-economic objectives of development policy

We have already emphasized the danger of a materialistic view in the ordinary meaning of the term. The time has now come for the nations to define precisely what objectives—apart from that of economic growth—they ought to lay down for themselves in their pursuit of development based on science.

We will pass rapidly over the military objectives. The United States of America, the U.S.S.R. and continental China aim at a balance of their retaliatory power; and this political objective, above all others, occupies a central place in their scientific planning. The aim of other nations is not to be dependent on a super-power for their supply of modern armaments. The argument at the present time, which cannot be developed here, hinges on the question of knowing whether a policy with this orientation leads to more—or less—international security for the nation concerned. It must however be mentioned, if only as a reminder, because of the place which this argument occupies in the science policy of a certain number of governments. In this case, we are dealing with 'a policy based on science' rather than with 'a policy of development based on science'.

The problem of the aims of space travel is more complex; it is one of the moot points of our age to explain what objectives are actually responsible for the allotment of such huge sums of money when the immediate gains, whether military or civil, appear to be marginal. Study indicates that space research is a complex subject in which military objectives, motives of prestige, economic motives, and aims of a fourth kind, to which we shall refer in a moment, are combined. We purposely speak of motivations rather than objectives, because the conquest of space appears to have its origins as much in collective psychology as in rational thought.

There is no doubt about the rationality of not leaving the field open for rival powers in a number of techniques which are very closely akin to those of armaments: propulsion, automation, interception, control and remote detection, etc. These are all techniques which have military application.

Concern with national prestige has so far been more subjective although it could be asserted that the U.S.S.R.—a country which, like Western Europe and Japan, is at the beginning of its second phase of industrialization—has enhanced its position in the world in a spectacular manner by rivalling, and at times even surpassing, in the sphere of space science and technology, a nation which is economically more advanced than itself. On this basis one could argue that concern for prestige is not an entirely irrational motivation. This point of view is sometimes also admitted—although with due caution—in business management.

With regard to the economic justifications of space budgets, the historian will no doubt say that they are examples of rationalization rather than rationality.

The stimulating effect of the space programmes on general progress has proved, in fact, to be far more important than could have been hoped by the men who decided to embark on them. We know also that, at the time, American industry felt an objective need for large national programmes, but without explicit expression. However, historical truth obliges us to say that these are *a posteriori* arguments, and that when the governments of the U.S.S.R. and the United States of America decided to finance space research, neither took this decision on the basis of its being a rational economic venture. Perhaps one day we shall know a little more about the part played by the subconscious, or intuition, in the decisions of governments...

However that may be, Europe and Japan, which are today embarking on these programmes in their turn, can consciously look for some profit from their expenditure on space, both in direct economic objectives (telecommunication satellites, or television) and in indirect results (the stimulation effect). But this was not the case before 1963 or 1964, the period in which the practical applications of the space programme were precisely fixed and the stimulation effect made itself felt.

There remains the fourth motive, which we have described as neither political nor economic: space as adventure.

Just as when the highest mountains were climbed and the poles were explored, so the ambition to overcome the worst difficulties, and by so doing to thrust back the frontiers of what man is capable of performing, is without any doubt the deepest motive of the engineers of the space programme and the astronauts. Is it shared by the governments which finance their efforts and their flights? It certainly finds an echo in the hearts and minds of all their fellow men, for to probe and thrust back the limits of man's possibilities is to give a rousing answer to the question which the whole species puts to itself: 'What am I and what are my powers?' From this it would be possible to put forward the hypothesis that the motive of adventure in space would remain—even at the level of governments—if the world scene were no longer dominated by the rivalry of the two super-powers whose main ambitions are for strength and prestige. It can certainly be doubted whether such huge budgets would still be devoted to space. We can also argue whether other more urgent needs ought not to be satisfied in the first place. But the promise which the United States of America and the U.S.S.R. have made to each other to co-operate 'after the Moon', so that they can achieve together the following stages in the conquest of space, gives some credence to this hypothesis that the fourth component of the decision to finance space travel is not necessarily the least important.

On this view, one can already speak of the existence of another objective of science policy, which would not be included in the ordinary aims of government of a political and economic nature. This objective would be the undertaking of great enterprises for the specific purpose of exalting man's life. Knowledge sought after for its own sake—an activity which, although it has never had large budgets at its command, has always remained alive in official apologias ever since the time of the royal patrons—would be a particular case of this, since we have seen that science is operational and that it is in action that man comes to understand the world and to know himself.

If such aims of research *have* to be classified, should not these be put first among those aims which have to do with the quality of life and the quality of society?—two ethical ends which every development policy ought to aim at, beyond economic growth and power, because they are the very goal of civilization. Will the reader allow such a bold classification of the aims of research? Perhaps he will, if he takes the view that this 'society of quality' would be nothing better than the most perfect of ant-hills, supposing it sought to limit men's ambitions to living a life in abundance and harmony —even a life made rich by the most profound knowledge and the most exquisite skills with all the refinements of culture—but sought at the same time to forbid the undertaking by mankind of all new and daring enterprise. In other words, can the solution of the problems arising from poverty, from the inadaptation of big cities or the deterioration of man's natural environment, or from the vicissitudes of community life, be considered as an ideal which, in itself alone, will satisfy man for ever?

Does not the 'life of quality' also mean that men must be allowed constantly to surpass their previous achievements by the exercise of their intelligence, will and courage; and that society should supply them with the means to do this in other ways than by war and the preparation of weapons of deterrence?

If the society of tomorrow succeeds in abolishing poverty, and if, in that society, it is only the great adventures of science which are in a position to offer man this 'new frontier', then he will indeed be justified in demanding that the necessary resources be made available.

But, as yet, the world in which we live is far from having become such a dull terrestrial paradise. The improvement of society in its various aspects— economic, social, cultural and human—is a task which has hardly been embarked upon. It is only two centuries ago since men began to think it possible. It is hardly twenty years ago that one nation, whose productive power was the most developed, emerged for the first time out of the age-old obsession with poverty, which still remains the lot of the majority of the people and will remain so until the end of this century. Before men can put their house in order, vast resources of intellect, will-power and courage will still need to be committed. We must now return to the qualitative aspects of this objective after this 'excursion' outside the bounds of rational motivation which the peculiar case of space travel forced on us.

From the quality of life, used this time in its practical meaning, arise above all problems of health, education, urbanization and the protection of the biosphere (nature, air, water).

From the quality of society are derived also the researches of sociology and political science, as well as scholarship preserving traditional or ancient cultures.

Let us for a short time return to these subjects, in order to emphasize their content and their significance, since this will make it possible later on to decide on the place which they deserve in a development policy based on science.

Medical research has no longer any need to be championed. This is not true of the technology of education, nor of town planning; their advancement is assuming an urgency which has unfortunately not yet been generally appreciated. Teaching, which today is at about the same point that medicine occupied a hundred years ago, must supply techniques for mass-education on a scale unprecedented in history. This effort must soon be complemented by an effort of similar magnitude in the field of adult education. This double demand, as we have shown, is the *sine qua non* condition of self-sustained and accelerated economic development. Although concern for education is to be found in the most advanced societies, chiefly because of the expansion of the universities and uncertainty about the security of an increasing number of jobs, it is in the developing countries that it will be most acutely felt. These countries should therefore logically take the lead in the scientific mutation of education, since they are experiencing the greatest pressure of needs.

The same argument applies to town planning. The population of the globe will double between now and the end of the century, and the whole of the population increase, or nearly all of it, will live in the cities. Therefore, half of the big cities of the year 2000 will be built after 1970. These figures give some idea of the immense possibilities open to applied science in a field where the achievements of modern society have remained very disappointing. But the picture of city development emerging from such a world average hides very different situations. In the countries which are already industrialized, the exodus from the rural areas is almost completed and the growth of the population has slowed down; their towns of today will only experience extension and rebuilding on a rather limited scale. In the developing countries, on the contrary, the largest part of the phenomenon of urbanization still belongs to the future. These countries can therefore still avoid the serious mistakes of the town planning of today. Consequently, it will also be their responsibility to give a lead to the many disciplines which together constitute the science of town planning.

With regard to the conservation and interpretation of traditional and ancient cultures, urgency springs from the desire to preserve the record of human achievement and values that development will obliterate or rapidly alter, and from the desire to ensure that the march of progress should disturb as little as possible the continuity and coherence of national cultures, as nations enter the path of accelerated development.

Conclusions on the choice of national goals

Leaving aside military and political aims, which do not belong, in the strict sense of the word, to the question of 'development based on science', we have stated that economic aims grow out of a development strategy, which itself depends on the phase of development which the country has reached. The objectives which are not economic are those concerned with the quality of life and society. The urgent need for research on some of these subjects (health, education, town planning, the biosphere in general, the preservation of the past) has been perceived and emphasized.

By this kind of thinking each government, after making a careful appreciation of the present situation of the country, as well as the situation which can be expected in the course of the next thirty years, will make a critical analysis of this probable future. From this analysis will spring the development blueprint and the corresponding political decision on national development planning. From the latter will be deduced the concrete objectives of national development over the next five, ten or twenty years. The lines of scientific expansion will then be chosen in order that these objectives can be fostered and attained. The growth of the scientific potential will sometimes appear as an end in itself, and at other times as a means to attain other ends. It will then find its place in an integrated plan of national development.

SURVEY OF A COUNTRY'S ACTUAL SITUATION

Between the choice of the general objectives of the national development plan and the precise definition of the resources and stages of development, there is a long intellectual process whose first two phases are the survey of a country's actual situation and the forecasting.

Aims and instruments of information

Science policy-makers must have accurate information on the current potentialities and resources of the country. They should also have available a clear analysis of the situation to which science is to be applied, and this must be as factual as possible. Third, they should work out how the situation would evolve if things follow their natural course, and should proceed to a critical analysis of the probable situation based on the current trends. This, then, is the starting-point for target-setting in science policy.

The information to be gathered is therefore wide in scope yet very specific. Furthermore, it must be collected on a permanent basis or, if not, at regular and sufficiently frequent intervals.

For these purposes, science policy-makers must have available powerful and flexible tools for data collecting, analysis and forecasting.

Survey of the scientific and technological potential

Evaluation of equipment and knowledge. There have been in the past attempts at such evaluations, but they were chiefly of a qualitative nature. These surveys contained descriptions of instruments which enabled the Academy of Sciences of Paris and the Royal Society of London, in particular, to attempt to assess the national heritage in the sciences. Or again, they were presented as 'state of the art' reports on science. Diderot's Encyclopedia is the most celebrated example, though not the oldest.

Survey of scientific activities. Before the Second World War, only the United States of America, the United Kingdom and, to a certain degree, the U.S.S.R. had any idea of their national expenditure in the field of research. These estimates, however, were not based on any special nation-wide survey.

It was not until these last ten years that governments realized that the act of expressing the scientific and technological potential of a country in quantitative terms was of significant importance. This came about when the resources devoted to research and experimental development reached a size large enough to attract attention in the national economic accounts.[1] It has also coincided with governmental awareness of the relationship between scientific activity and economic development.

Many governments have understood the profit which they could derive from systematic and detailed information about national research activities. In fact this information allows them: (a) to measure the relative strength—in terms of resources—of research centres and research teams, and to assess the maladjustments and imbalances which may exist, as between the disciplines, as between orientations and as between geographical regions, etc., or again in the distribution of the nation's scientific and technological potential (units that are too numerous and possibly too small, or on the contrary an over-concentration presenting the danger of rigidity and failure in communication); (b) to establish, by means of periodic surveys, chronological series of data; these series make it possible to make projections concerning the future evolution and orientation of scientific activities; (c) to draw up, by combining these data with other quantitative data—particularly economic data—indicators of scientific activity, bearing particularly on the relation between expenditure on research and the national product, the added value in productive activities, or the investments in the various industrial sectors; similarly, personnel employed in scientific activities can be compared with the working population, with total personnel in the various economic sectors, branches of industry or universities, etc.; (d) finally, to compare the national scientific effort, and its organization, with that of foreign countries.

This last point requires a few comments.

The comparison of national scientific efforts

Because of the present insufficiency of 'internal' criteria which would make it possible to evaluate the intrinsic usefulness of a particular piece of research, and of criteria which make it possible to measure the economic effects of that research, governments are making use of the method of comparisons, in order to assess their scientific situation. They are seeking to establish the differences which exist between their own scientific potential and that of other countries which have reached a comparable level of economic and social development. By a systematic analysis of these differences, they try to measure their deficiencies in science, and thus to assess the extra effort which they have to make in order to put themselves on a level with their competitors.

A similar method can also be useful for those countries which wish to pass over into a new stage of development by applying a purposeful science policy. In this case, their scientific and technological potential must be compared, not only with that of a country in the same group, but also to that of countries which have attained a more developed economic and social organization than their own. Thus a typology of successive stages of develop-

---

1. It is usually at the time when the resources allotted to science reach the threshold of 1 per cent of the GNP that science policy comes to the forefront of the national scene.

ment, which has only been outlined in this essay, is steadily emerging. Such a typology ought to make it possible for each country to find its proper place, and to identify the fields in which it needs priority of effort.

This kind of approach seems to be spreading today. It is being adopted not only by the scientifically less developed countries, but also by the European countries, which are directing their attention to the United States of America in order to estimate what their over-all level of scientific effort ought to be. The United States, for its part, does not hesitate to admit that in certain fields its scientific activities show deficiencies in comparison with those of Europe. Above all, this method makes it possible for a country to find its proper place by reference to the world context.[1]

Standardization of methods of evaluation for comparative purposes

The usefulness of such comparisons has called for important work to be done on an international plane. The aim was to establish, by mutual agreement between nations, a standardized system of definitions and procedures for the collection of statistics on research and education, and so make it possible to compile information on national potentials and their orientations, from which comparisons can be made. In this way the collations which have been made within the framework of OECD have made it possible to establish standard guidelines for the collection of science statistics, on the basis of which a survey was carried out in all OECD Member States under a programme entitled International Statistical Year of Research and Development. The breakdown of the results has provided a certain number of nations with the first opportunity to compare their respective levels of scientific effort.

Unesco has put on the agenda of the meetings of experts and the conferences of Ministers of Science, which the Organization calls periodically in the different regions of the world, the arrangements that should be made in order to extend the possibility of such comparisons to all nations.

In particular, Unesco has launched in 1968 a statistical survey of the research effort of all its Member States in the European region.

This survey is based on a standard questionnaire which makes it possible to reconcile and compare the science statistics collected by European countries under the aegis of OECD on the one hand and of CMEA (Comecon) on the other. The results of the investigations have been examined by the 1970 Conference of Ministers responsible for science policy in the European Member States of Unesco.

The establishment of a standardized method for the collection of science statistics has given rise to a number of technical difficulties, the most significant of which concern the very definition of the activity of research as part of the scientific activities of a country.

What emerged from the attempt to discover just how much was being spent on research, and just how many people were in fact engaged on research, was the realization that the very concept of 'scientific activity'—

---

1. However, it cannot be given a normative value, since it is well known that comparison is no proof.

which had hitherto appeared to be unambiguous—was in reality ill defined and covered several types of activities. Until recently these definitions had remained on a strictly axiomatic plane: they could indeed have remained so, as long as research remained an isolated, almost individual activity, which in any case was largely unorganized. But to comprehend research statistically, it became necessary to recognize that the idea of 'scientific activity' was in fact a concept that was at the same time wider in scope and more complex than that which had been implicitly accepted before.

In fact, science statistics cover an ever-widening range of scientific and technical activities. Besides research (fundamental and applied), experimental development, and a series of related activities among which can be mentioned certain aspects of education and training touching upon science and technology, these statistics also include the scientific public services and a whole series of activities which form an integral part of the process of production of goods and services (for example, laboratory testing in factories, prospecting for mineral deposits and oil, etc.).[1]

The earlier axiomatic definitions have consequently been replaced by definitions based on genuine functional analyses.

Science statistics, the last-born of the branches of statistics, often still encounters psychological barriers. Scientific authorities in universities and in industry are sometimes reluctant to co-operate in any detailed surveying of their scientific activities, either because they do not understand the real meaning and scope of the statistical method or simply because of a defensive reflex. Some of them still seem to think that a statistical survey is only the prelude to a more thorough inquiry which is intended to set the fashion for arbitrary intervention by the government in their internal affairs. No doubt the meddlesome and heavy-handed efforts of some government administrations have contributed in the past to create this strong prejudice that 'research activity cannot be expressed in figures'.[2] As the first surveys of the national scientific and technological potential are completed and their analyses published, it is likely that these prejudices will fade away.

Ill advised government interference is not the only ground for fear. Competition is another, at least as important, which affects industrial companies.[3] Resistance to statistical surveys also affects bodies whose non-profit-making status ought to sway them more readily in the direction of co-operation.

---

1. *Manual for Surveying National Scientific and Technological Potential*, Paris, Unesco, 1970. *Questionnaire on Statistics of Research and Experimental Development Effort, 1967*, Paris, Unesco, 1968.
2. This reaction appears to be more common in Europe than in the United States of America, because in the latter grants and contracts for research allocated by the Federal Government form a powerful motivation for making returns in statistical surveys.
3. A striking example of this is the case of the Netherlands. The five largest industrial firms (which are business groups of international importance, possessing an almost complete monopoly in their respective industrial sectors at the national level) are opposed to the separate publication of the figures for the expenditure on, and personnel engaged in, research in these industrial sectors, because this action would make their efforts too easily identifiable.

If the principle of the survey of scientific and technological potential, on which the preparation of science statistics is based, has now been everywhere admitted, the scope and the frequency of these surveys varies from country to country. In particular, science statistics are most often confined to the manpower employed in scientific activities and to the expenditure allotted to these activities. The surveys do not often include a record of scientific equipment, and rarely the numbers and objectives of the research projects at the level of the research units where the activity is developing.

We may hope that the efforts made to set up a standardized system under the aegis of Unesco[1] will overcome much of the resistance encountered hitherto, and lead to making these surveys of the scientific and technological potential of the nations more regular and more complete.

An indispensable extension of the inventory: monitoring the results of research

At present, surveys of scientific and technological potential deal with the human and material resources devoted to research, e.g. the inputs. But as yet the results of research—that is, the outputs—are not measured, even on the economic plane.

Since the problem of political choice is presented in terms of effectiveness, study of the output of research activities, and the collating of these results with the resources employed, will become increasingly important.

This subject has so far remained the preserve of individual economists. Although the literature devoted to it is already quite extensive[2] nothing is yet discernible which could form the groundwork of a general theory about the relation between research and economic development.

On the level of the choice of research projects which are to be financed by a government or industrial enterprise, the assessment of the probable results is now making progress because of several applications to projects of methods of cost effectiveness analysis, invented to evaluate investment. What is invariably missing is the evaluation *ex post facto,* and the critical examination of past decisions retrospectively.

FORECASTING

Scope and nature

The surveys dealt with above provide a description and an assessment of the scientific and technological potential, covering the recent past and the present. The object of forecasting is to describe and estimate this potential in future, but this time by projecting its evolution.

If it is correctly set out, a survey of the national scientific and technological potential mirrors the present situation. Forecasting, on the other hand,

1. *Manual for Surveying National Scientific and Technological Potential.* Paris, Unesco, 1970.
2. See the final bibliography of: *Recherche et Croissance Economique,* II, Brussels CNPS, 1968.

foretells possible or probable situations. By definition, it does not lead to definitive conclusions.

Forecasting derives from hypotheses on the orientation and amplitude of the change. The simplest and most usual forecast is that of projecting past trends. Other forecasting methods rest on hypotheses derived from theoretical or econometric models which are devised to reflect the internal dynamics of the system.

Forecasting cannot take any account of exceptional events which might modify evolution. It cannot therefore build or imagine the future that would ensue from data which were outside the expected course of evolution.

Since it is thus confined in its scope, forecasting does not contain any normative elements. Its results can in no way be considered as objectives. Like the survey of scientific and technological potential, it provides part of the information at the disposal of decision-makers. It foretells the probable condition of the system on the hypothesis that no action will be undertaken to modify the present course of events.

It shows in particular how the scientific potential would develop if it were left solely to the spontaneous dynamism (or inertia) of its driving forces, or if government support remained at a constant level.

Research, like many activities, shows a tendency to grow under its own momentum. This is particularly true in the case of projects suggested by the research workers themselves, which are largely the reflection of earlier policy decisions. It is only natural that teams of scientists, specializing in nuclear physics or agronomy, will suggest, for government aid, projects in those fields. In the absence of guidelines emanating from the science planning authorities, the distribution of grants between the various orientations of research will be largely a reflection of the capabilities previously deployed, with an emphasis on those that show the greatest vitality, no matter whether the vitality is that of the scientific field in question or that of the organizational set-up for research in that area.

The study of such internal dynamics of scientific research has given forecasting-by-extrapolation more than marginal significance.

Field of action, and usefulness

Forecasting serves to identify features of evolution which can or ought to be accepted, and also those which ought to be corrected.

Two examples will make it easier to understand this.

First, let us assume that in a certain country a forecast suggests that one must expect the number of students at the universities to double in fifteen years. In face of such a forecast, three attitudes are possible, from the quantitative point of view, for those responsible for science policy: to 'accept' this evolution, or to decide to 'slow it down', or to decide to 'speed it up'. In the first case, they know that arrangements will have to be made to accommodate the students expected; in the second, they know that measures will have to be taken to ensure a more stringent selection of the candidates and to channel the surplus towards a different kind of third-level education; in the third case they know that secondary education must be developed, access to education facilitated, the number of grants increased etc. ... Moreover, in the first and third cases it will naturally also be necessary to adapt the resources and the teaching staff of the universities to the expected size of their intake. Decisions can be made on the type of measures to be taken after having compared the probable situation—a doubling of the number of students—with the situation that is desirable.

Second, let us now assume that in the same country the forecast suggests, for the same period, a more than average development of agricultural research and a relative slowing-down of advanced technological research. Here too there are two possible kinds of reaction on the part of the government authorities. According to whether this evolution is judged as being either contrary to the strategy adopted for development (during the second phase of industrialization), or else in conformity with that strategy (during the first phase), so the forecasting will either call for corrective action or not, by the policy-makers.

These two examples, which are intentionally oversimplified, nevertheless entail some additional comments.

Forecasting can only be useful for decision-making if it includes a sufficient number of significant features of the present and future situation. Thus, the simple indication that the number of students could double in fifteen years offers virtually nothing of interest, unless it is possible, at the same time, to get a clear insight into the foreseeable evolution of the demand for university graduates, and to compare the two figures in the light of the general needs that arise from the over-all national development plan.

In the same way, it is simply of no interest to know that aid for agricultural research is developing faster than industrial research, unless one has accurate information about the actual rhythm of progress of agricultural productivity, or about the decrease of the working population in agriculture, etc.

For this reason scientific and technological forecasting cannot be separated from general economic and social forecasting. In fact, thinking concerning the objectives will rather rely on over-all development forecasts, whereas thinking on the use of science for application will tend to rely on the forecast relating to the growth of the research and experimental development system itself.

The corollary is, of course, that cross-checking of the forecasts made in different spheres or economic sectors is necessary, in order to ensure that these are consistent, complementary and compatible with each other in their scope, nature, methods and hypotheses.

It is by introducing limiting factors of consistency that one can correct the cursoriness which often characterizes the simple projection of tendencies. As an example, a forecast of the number of university students ought to be compatible with the forecast of the numbers of persons completing secondary education, and these in turn ought to be compatible with the corresponding age-group of the population. The latter forecast obviously depends on the natural growth of the population, a non-linear phenomenon, and the forecast of the balance between emigration and immigration, among whose principal parameters is the level of the GNP. Obviously by thus tracing back the successive causes of the phenomenon, one obtains a firmer forecast than by simply extrapolating the curve of the numbers that enrol for university courses.

The more generalized the forecast is, the more it gains in value. But the number of parameters and the number of the equations which link them increases rapidly. The calculations quickly become too complex and take up too much time, especially if one is forced to update them frequently. One is thus led to work out by computer a complex national projection, so as to achieve automatically the consistency and updating of the various forecasts.

Sooner or later, in the countries which decide to apply a development policy based on science, the need for such a methodology becomes apparent.

Later improvements

To forecast how a situation will evolve in line with the present trends is, as we have said, a very useful operation. At a later stage, however, it will become necessary for forecasting, while preserving its informational value, to succeed in indicating to science policy-makers the situations which are likely to occur if they take one decision rather than another. Forecasting then becomes simulation.

At first sight this process hardly appears more complicated than the former one. But this is not so, since from the very moment that one abandons the hypothesis of the simple extrapolation of present trends, one introduces into forecasting dynamic data whose interdependence is much more difficult to unravel and measure. It is necessary to identify the variables and to determine with accuracy what their functional relations are. These relations take the shape of a system of equations or a model whose variables and parameters have to be calculated or estimated. It is one of the objects of econometry to provide mathematical instruments for forecasting and simulation.

Recent publications of Unesco recommend the general adoption of these methods.[1]

Future-oriented thinking in science and technology

So far we have dealt only with quantitative forecasting. Qualitative change has not been taken into account, such as a significant discovery, a technological breakthrough or a new scientific theory. Such a forecast of scientific development made in 1940 would have allowed no appropriate place for nuclear technology, nor for the technology of computers or even of data processing. This example shows that purely quantitative forecasting for twenty or thirty years ahead loses much of its relevance, because the scientific and technological breakthroughs—a process akin to mutation—cannot be foretold by the extrapolation of former trends. One cannot hope to foresee all such mutations, but there are some which can be foreseen and even

1. Unesco, *Structural and Operational Schemes of National Science Policy* (Science Policy Studies and Documents, 6), 1967, p. 22-3.

expected by shrewd and observant experts. This expertise should be utilized by the planner.

Future-oriented thinking in science and technology is not limited to the forecasting of discoveries and inventions. It goes even farther than the forecasting of the products which will spring from future inventions.[1] As it is already practised in certain countries, it aims in fact at anticipating changes in the structural organization of industrial production caused by the mutations of technology, and, in a more general way, at anticipating the economic and social effects of forthcoming technological advances.

In its present state of development, such future-oriented thinking covers not only industrial technology but also all the theoretical elements—basic knowledge, techniques and processes—which are associated with technological progress and which find expression in the management sciences and data processing. Its scope is necessarily very wide.

By looking well ahead, future-oriented thinking in science and technology makes it possible to distinguish certain relationships between fundamental research and applied research. It is useful when trying to estimate the amount of fundamental research that should be performed in order to keep programmes of technological research productive in the long term, and to prevent them from getting bogged down in unpromising avenues.

An example will illustrate this role. Controlled nuclear fusion is expected to take over, at about the end of the century, from fast reactors for the production of energy; commercial development of this latter process is expected in the eighties. But the take-over by nuclear fusion depends upon the overcoming of serious technological difficulties. The experiments undertaken up to the present have failed through lack of sufficient knowledge of ultra-high-temperature plasma physics. It is known therefore that an important phase of successful basic research must necessarily precede a technological breakthrough. Future-oriented thinking in science and technology must identify the problems which have not been resolved, which are hampering development today and could turn out to be bottlenecks later on. Further, it is necessary to identify the auxiliary branches of technology in which progress will be necessary before the breakthrough can be made (for example, the development of materials which are resistant to higher levels of temperature or irradiation than the materials of today).

Future-oriented thinking in science and technology aims also at defining the conditions in which new organizations could be created whose mission would no longer be exclusively oriented towards the marketing of special products, but also towards the fulfilling of various other functions (aiming at collective objectives and multipurpose missions) within the framework of integrated systems rather than of the economic sectors which we have today.

In this context, future-oriented thinking helps to estimate the likelihood and the rhythm of the substitution[2] of one product for another or one tech-

1. See especially on this subject: E. Jantsch, *Technological Forecasting in Perspective,* Paris, OECD, 1967.
2. Substitution of oil for coal, and atomic energy for oil, is an example in the economics of energy production. The transport sector is also subject to frequent substitutions, which should henceforth be planned in terms of integrated transportation

nology for another. This is the most delicate problem for long-term economic forecasting.

Such forecasting suggests various alternatives, and aims therefore at facilitating the choice between the different ways of producing the same effects or attaining the same objective.

To sum up, future-oriented thinking in science and technology makes it possible to enlarge the space-time dimensions of a science-based development policy, and allows for a synthesis to be made between economic policy and science policy.

BUDGETARY ANALYSIS AND PREPARATION OF THE BUDGET

Need for analysis

Budgeting for scientific activities is, in every country, the concern of various departments or offices of government.

Until the recent past, each ministry or office wrote into its budget the credits for scientific activities which it deemed necessary to support action in its own sphere of competence. Hence, no over-all government programme and budget existed for scientific activities.

When government expenditure on higher education and scientific research began to reach significant levels, the necessity was felt of avoiding excessive dispersion of effort by introducing greater consistency in budgetary appropriations for science and research. This called for a new kind of budget preparation. The first stage towards a reform of this kind was the analysis of the appropriations ascribed to 'scientific activities'.

Scope of analysis

The nature and detail of the information which ought to be supplied for each appropriation depends on the use one intends to make of the analysis. In any case, analysis must deal with both the appropriations allowed in the previous financial years and the appropriations requested for the next financial year. The observed differential growth of the appropriations is, in fact, one of the most fruitful results of an analysis. In fact, analysis can be considered either as a means of finding out about government action in scientific matters, or as a tool of the government's science policy.

requirements. The collective failure of the private car has made the necessity for a substitute obvious to all. But one cannot study the best way of getting people from one place to another and the associated problems of cost and flows, without examining the users' motivations. This will make it possible to isolate the innumerable cases when people go from one place to another to communicate with one or a group of other persons. One ought therefore to study also new possibilities of substituting telecommunications for the transport of people. First, the postal service and then the telephone have already made many journeys unnecessary. In future, this substitution could be fostered by multiple television conversations which would replace meetings, as soon as the price of this type of communication becomes competitive with the cost of transportation. Here, future-oriented thinking links up with cost-effectiveness analysis.

147

A means of monitoring government action in the field of science and techno-logy. Budgetary analysis can confine itself to the basic information which is supplied by the accounts of the government departments, offices or agencies. It can also aim at a more exact and detailed knowledge of the use of government funds: (a) by splitting up every appropriation which covers both scientific expenditure and expenditure of another kind; (b) by seeking to retrace the path of the government money through the national system for science and technology right to its final end, which is the effective expenditure; (c) by making a distinction in the reporting between regular expenditure and new expenditure; (d) by making a distinction between intramural expenditure and extramural expenditure or transfers of money in the form of subventions or contracts; (e) by itemizing the nature of the intramural expenditure into personal costs, capital costs, running costs, etc.; and (f) by defining the function to which the expenditure is related (research, education, government service, logistics, etc.).

If it is renewed every year, such analysis grows more comprehensive every time and assembles the facts which will make it possible to assess the impact of government action on national scientific activities and on the economy of the country. At this stage, the results of budgetary analysis must necessarily be compared with the information supplied by the survey of the national scientific and technological potential, both government and private. This comparison allows some interesting cross-checks to be made.

The difficulties of the analysis. When limited to a simple compilation of the appropriations foreseen in each department for scientific activities, and to a classification according to their original stated purpose, budgetary analysis does not run into serious difficulties. Such a task can usually be carried out by using the wording of the different headings figuring in the departmental budgets which appear in the over-all State budget, together with their explanatory commentaries. But difficulties arise when one wishes to classify the appropriations according to their final destination (recipient organizations or recipient persons), or again, according to the exact nature of the scientific activities which these appropriations will enable to be efficiently financed (education, fundamental research, applied research, experimental development, etc.).

There are usually a number of stages in the financing of research by a government. These stages sometimes take the shape of intermediate agencies for the distribution of appropriations which possess a greater or lesser autonomy. The intervention of these agencies (and the exercise of such autonomy) often changes the original destination of the appropriations.

Furthermore, the traditional classifications of government accounts are insufficient to embrace scientific activities, and even less adequate to enable functional distinctions to be made between the appropriations. The difficulty resides chiefly in the fact that a certain number of appropriations for research are hard to dissociate from the other functions of the ministerial departments. Besides, there can be a certain amount of resistance, if the description of the appropriations has to be drawn too finely.

148

With the purpose of helping countries to overcome such difficulties, a certain amount of theoretical research has already been conducted on this subject, but much more still needs to be done. It seems indeed that the best method consists in beginning by simple budgetary analyses and making these progressively more detailed and more searching as time goes by.

Utilization of budgetary analyses. The analysis of budgets serves first as a normative guide to the channelling of government funds for science and technology and to the management thereof. Let us suppose, for example, that it has been proved that allocations of the same nature and for the same purpose—such as doctoral fellowships—have been made by several administrative bodies according to different principles, although they are identical in aim and are meant for the same category of candidates. It may happen that some administrative bodies are also charging other expenses to the budget earmarked for doctoral fellowships.

If rivalry springs up between administrative bodies for the granting of fellowships, the result can be an attempt to outbid each other, and a corresponding lowering of standards may occur, while other activities, less open to competition, are neglected. It may also happen that in such a multichannel financial system, some sociological groups make application to one body rather than another, so that there springs up in the administration and financing of science a system of 'clients', in the Roman meaning of the term, which in turn generates pressure groups and genuine power structures.

Normative action in the case of doctoral fellowships might consist in not having more than one financial channel for such a purpose, and in defining the corresponding budget appropriation in such a way that it cannot be used otherwise. Generally speaking, it would then be possible to regulate the volume of the resources allotted to a specific function by fixing the amount of the corresponding annual appropriation. In the case of fellowships, it would also be possible to ensure that identical rules of priority, selection, calculation of the grants, etc., were applied to all the candidates. Later, the appropriation for doctoral fellowships could be made more specific, in order to fit in with over-all science policy decisions aiming at strengthening one discipline or another.

Budgetary analysis can also, by means of the data which it assembles about regular expenditure and contingency allocations, help to define the margin of freedom which the government has available to launch new enterprises and to decide on the over-all balances required for the development of science and for the national economy—decisions as between the different categories of expenditure, between the kinds of organization which need support, between the fields of science which need to be promoted, etc.

Furthermore, cost analysis makes it possible to forecast expenditure, and so facilitates advanced preparation of policy decisions. Finally it makes it possible to integrate science policy in the general policy of the government, particularly by providing a means of ensuring that the growth of government appropriations for science and technology is taking a normal course in comparison with the evolution of total government resources.

149

Planning through the budgeting system. The principal use of budgetary analysis is, however, the preparation of planning as such.

The planner is not a manager. He must not bear responsibility for the management of institutions or contracts, because he would lose the detachment which is necessary to keep a clear judgement on the advisability of expenditures. But, equally, government operations must not by-pass the planner. He would then become merely a consultant or a shrewd observer of government action, responsible merely for analysing the situation periodically and putting forward some quantified objectives. This stage of advice and recommendations is perhaps a necessary first step in the planning process, but it belongs more to the exhortation and verbal exercises known as *wishful targeting* than to actual development planning.

In the capitalist and mixed-economy countries, the means by which the planner influences action is the budget. The Act of Parliament (or of the Executive) by which a specific sum of money is placed at the disposal of a Minister or a government institution for the exercise of a given function is, in fact, a major decision. Such a decision determines, for the Minister or the institution, the boundaries of their freedom of action, since the object and amount of the expenditure are specifically defined. The recipient nevertheless retains entire responsibility for the way in which the money will be spent. It is by co-ordinating the entries in the budget—the main task of 'macro'-decision-making—that a head of government turns the whole of the public sector into an operational system geared to an integrated and coherent governmental plan of action. At a later stage, the whole nation may come under the catalytic effect of this all-embracing plan of action, since the impact of government expenditure nowadays influences every sphere of life. The budget is therefore the principal instrument by which the consistency of the national development plan is impressed on all parts of a government and, by implication, on a nation.

It is clear that this consistency involves the whole government, and that the elaboration of the budget is by nature an interministerial activity. It is through the formulation of the budget that it becomes possible to ensure the convergence of the action of the various Ministries, without the head of the government having either to interfere with the management responsibilities of each of his colleagues, or to intervene afterwards in their 'micro'-decisions.

In short, public finance, in this respect, evolves in the same way as the finances of private enterprises, since the introduction of budgetary planning. This new method has changed the original function of budgeting. It is no longer only an act of authorization of expenditure. It becomes an act of adaptation and optimization of the expenditure with reference to the objectives to be attained. The unit of reference is no longer the administrative set-up, but rather the function that has to be carried out—or, more precisely, the mission, understanding by the word 'mission' a group of functions which unite to attain the same objective.[1]

1. It is particularly the aim of the Planning, Programming, Budgeting System which has been put into general use in the Federal Agencies in the United States of America.

If one were to make a comparison with biology, it could be said that the traditional budget was invented to ensure the vegetative functions of the government and was intended, in consequence, for monitoring the continual renewal of the same expenditure. Since modern governments are action-oriented, their budgets should be planned accordingly. Action consists in carrying out successive missions. The expenditure ceases when the mission is fulfilled. Other missions, aiming possibly at other goals, take its place.

One cannot indeed replace the administrative body (Ministry or organization) as a basis for the classification of appropriations, since the responsibility for spending the sum entered in the budget will lie with the head of that body. But in order to fix the sum to be entered, it is first necessary to have regrouped all the expenditure contributing to the same mission, whatever administrative body may be entrusted with the duty of managing it, and next to examine the needs of each body in the light of the total expenditure needed by the mission as a whole. Then, within this new grouping, will be isolated the functions which need the joint action of several Ministries or organizations in order to ensure that the sums necessary for their achievement are entered in the budget of each ministerial department involved.

Thus, as the advancement of science is a general mission of the government, the appropriations for scientific and technological activities are worked out in a 'science budget'[1] before they are entered in the budgets of the various Ministries and government organizations. After they have been worked out at a second stage, a regrouping by missions and functions will be made. There will be, for example, the plan for the development of nuclear energy, the plan for the expansion of the universities, the plan for the development of technological research, etc. The sum of these projects comprises the governmental science policy in its proper sense. It is by studying them that thinking progresses from the definition of the objectives to the calculation of the necessary resources, and to the administrative authorization of the expenditure. The same regroupings and the same stages of thinking are necessary to evaluate *a posteriori* the effectiveness of the expenditure, an evaluation which leads to optimization by feedback.

The process of preparing for the expenditure and the assessment of its need thus takes considerably longer than it used to do in the old days. But the process pays in the end. First of all there is the 'financial pay-off': by simple subtraction, items of expenditure are uncovered which do not belong to any 'mission' actually in operation. More often than not these will be cases of survival from past missions. If it can be shown that this is not a permanent and necessary function, deriving from a 'vegetative' function, this expenditure will be struck out of the budget. Another worth-while side-effect shows up when the administrative and managerial staffs of the various Ministries, etc., draw up integrated horizontal budgets and learn to work together to draw up a quantified plan of action for a specific government mission in which they have different parts to play. The budgetary planning procedure is therefore also an educational one; it removes the barriers between administrative bodies by teaching each one to make use,

---

1. In certain cases this budget includes all or part of the budget for higher education.

for its own practical ends, of the ability and machinery which exist in other parts of the government. Procedures are thus developed by which the various departments can in fact count on each other, since each department has the possibility to have entered in the budgets of other departments the sums necessary for those latter departments to accomplish certain specific tasks, which the first department needs for the complementing of its own action.

The first stage on this road, however, presents a difficult hurdle to surmount. It must be seen to that each Ministry or organization inscribes in its own budget a heading or section labelled 'Science'.[1] Next, this section must be detached at the time of the vertical discussion between the Ministry of Finance (Ministry of Planning, or Bureau of the Budget) and the Ministry in question, so that the 'horizontal' discussion of the appropriations of all the science sections under consideration, regrouped temporarily in a National Science Budget,[1] should be held simultaneously at the highest level of government in the presence of representatives of the department for science planning.

The total National Science Budget[1]—that is, the maximum total amount of this temporary 'horizontal' budget—will first be fixed by a ministerial committee for science policy, which will later on hold discussions about the total National Science Budget as a whole, and the individual allocations which it includes. So with their general outlook already decided upon at government level, the sections labelled 'Science'[1] will take their place again in the budgets of the various Ministries.

In short, the object of this process is to introduce in one area of government action, namely science policy, the prerequisite stages of formulation by missions and functions which ought to provide the basis of the whole budgetary planning procedure of the government, if its organization were totally development-oriented.

Of course the governments which have already adopted a generalized system of budgetary planning will dispense with this stage. But other governments should certainly not wait for the general reform of the budgetary planning system to set up a procedure for the annual preparation of the national science policy budget.

CONSULTATION AND CONCERTED ACTION

Definitions

Consultation and concerted action are the procedures by which the government seeks the opinions or the joint action of individuals and organized bodies which are outside the machinery of State.

We shall speak of *consultation* when the government asks the opinions of outside persons about its own affairs and about its own decisions. Such advice does not bind a government.

1. Or, according to the circumstances, 'Science and higher education'.

On the other hand, we shall speak of *concerted action*[1] when the persons or outside bodies consulted have to take certain decisions for action on their own behalf, and when it is desirable that they should form an agreement with the government so that both their decisions and those of the government should be made compatible and complementary in the best interest of the national objectives. Thus, decisions taken about science policy by the universities (whether they are private institutions or autonomous public corporations) and decisions made by large industrial enterprises (which can also be privately owned or publicly owned and autonomous) must be in harmony[1] with those of the government. They are the subject of concerted action. Where there has been general agreement, all the parties ought in principle to be bound by the joint decision.

Evolution of consultation

Whenever it has lacked the requisite expertise in persons among its own ranks, a government has always had to take advice and counsel from outside in dealing with scientific and technological problems. Simon Stévin, who was in the service of William the Silent, was truly a scientific adviser to his sovereign. The Committee of Scholars, who gave advice to the Convention of 1792, can also be seen as a prototype of the scientific advisory councils set up by the governments of today.

However, during the last twenty-five years two important evolutions have taken place.

First, the function of advising governments on scientific and technical matters has become institutionalized. Thus, in the sphere of the promotion of research, the specialized organizations responsible for distributing State grants have set up scientific committees of independent experts and have asked for their recommendations on the scientific value of research programmes put before them. Also civil and military administrative bodies, who allot contracts in order to achieve the objectives for which they are responsible, have formed the habit of consulting panels of scientists on the over-all strategy according to which research ought to be developed and, at a more operational level, on the detailed objectives of programmes and the quality of research teams or contracting companies. It even happens that government research establishments have permanent scientific councils, composed of specialists who are chosen partly from outside the government.

Secondly, in certain cases, the aims of the consultative function have been enlarged; several countries have recently set up councils at the urgent request of the government, whose duty it is to give their recommendations on the over-all national science policy (both as regards the promotion of science, and science-based development).

---

1. It should be made clear that *concerted action* is here used to refer only to agreement between persons or services not subject to any common discipline. *Co-ordination*, on the other hand, is used to refer exclusively to the alignment of persons, or of services, which are subject to such a common discipline; for to co-ordinate, one must have the authority to draw up an order.

One must be careful not to confuse these science policy councils with the scientific councils described above, which examine the advisability of particular courses of action. Even less must they be confused with 'national research councils' which are actually councils for the management of the national organizations for the promotion of scientific research and which, for this reason, are invested with the responsibilities of decision and even of the management of funds.

The frequent incompatibility of the responsibilities of management with the consultative function has not always been perceived.

We have already pointed out that certain legislators or certain governments have thus combined the making of recommendations on science policy with the other functions of the research councils, in particular that of distributing funds. This additional science-policy-making function could usually not be carried out properly. For how could they make a sound judgement on the use which the other agencies make of government resources, or the best distribution of these resources between various agencies, when they are themselves utilizers and managers of part of the funds in question?

Scope and results of consultation

Governments have to solve numerous problems every day which derive from their own responsibilities alone. If they consider that a contribution from science is essential for the better solution of some of their problems, they usually decide to 'consult' appropriate specialists. But in these cases there is no 'concerted action', because the decision which has to be taken depends on the government and nobody else.

When therefore it is solely a matter of taking advice, it is the rule that such advice does not bind the person or the authority asking for it.[1] It is in the interest of those who request advice that the consultants should be as competent and as independent as possible. If the advice of a council is sought, its members will be appointed in their personal capacities and not because of their functions or as representatives of non-governmental organizations.

Scope and results of concerted action

When, on the other hand, one has to define a national development policy based on science, a government is no longer the only actor in the play. A number of actions have to be made to fit in with each other, and this requires that those in control of the various participating bodies co-operate in the formulation of joint objectives and operational procedures. The technique of 'concerted action' is a response to this necessity. It will be noted that this technique can also be applied at other levels than that of the planning of national science policy. This is the case when the government wishes to harmonize details of its actions in a special sector of the national

---

1. There are certainly cases where the advice must of necessity be followed, but these are exceptional; administrative law has formulated detailed considerations on this subject, under the concept of 'joint responsibility'.

activity with those of organized groups and non-governmental organizations. In this case the government will also enter into concerted action with them, but the objectives and the methods of procedure will be less general and more detailed, that is, their plan of action will be more expressly operational.

To what extent does a 'concerted action procedure' bind the parties? Since this is a new technique the general principles of law are of little use on this subject. In fact the problem raised here is whether plans and programmes formulated in this way have a binding force or are merely guiding principles.

Theoretically several solutions are possible.

First one can imagine that after the partners have agreed to collaborate, each one retains the freedom, in his own sphere, to take the decisions which he deems necessary, useful or simply possible. In this case the procedure will only have provided guiding principles for all the partners, including the government. Its purpose will have been: better exchange of information, a joint effort of reflection on the problem, and the elaboration of a common viewpoint of the general line of conduct to be adopted. This arrangement is the one most usually employed today. It is the source of frequent disappointments, for each partner has a natural tendency to think later that the others have not acted according to what had been jointly proposed for action.

In a second type of arrangement one can imagine that the results of the procedure are binding for government departments, while maintaining the character of guiding principles for the other partners. Such understanding, which is a kind of tribute paid to the activating role of the State, runs the risk of being deceptive and inappropriate in practice. The State possesses in fact the sovereignty which allows it to disengage itself, in a completely legal manner, from unilateral commitments that it may have taken. Besides, if the other partners avoid taking the decisions which devolve upon them, governmental action is in danger of falling flat and producing insignificant results.

Finally, in a third arrangement, one can imagine that the results of the procedure are binding on all parties, each one being responsible to the others for the commitments he has undertaken. These actually represent obligations of a special kind, like those which result from collective bargainings between trade unions and federations of employers. Although they may not be sanctioned by a court of justice, they can have a powerful binding force, if the need for co-operation is strongly felt.

This last arrangement would, in principle, produce the most effective science policy. However, it comes up against numerous difficulties, including that of bringing together in an 'organ for concerted action' the people who have the actual power to pledge the support of the groups which they represent, and the necessary authority to have these undertakings carried out by those groups.

For this reason it is open to doubt whether an agreement at the summit, having a bearing on the whole of a policy, would bring results at first. But more definite agreements could be concluded in instalments, in the shape of development contracts with universities and industrial enterprises which make an appeal for public funds. The negotiation of these contractual arrangements could give rise to a far-ranging debate on the long-term strategy of these independent bodies. True, by struggling to hang on to support which is automatic and unconditional (in the form of block grants) they have often sought to insure in advance against the obligations of concerted action. They prefer that 'concerted action' remain a platonic concept; in it, they find a pretext for blaming the government without, in return, any obligation to give an account of their own plans for the future and their own management.

Various methods of organizing concerted action

Should concerted action take place within one single organ uniting the government and all its partners? Should these partners be chosen in their own right, because of their qualifications and personal influence, or should they be called upon in their capacity of responsible representatives of the groups whose spokesmen they are invited to become?

Different solutions are in use today.

In the United States of America, a country traditionally attached to the principle of checks and balances, the President has dealings with a large number of organs, some of which are composed of scientists, others of representatives of the business community, and still others of members of the administration and government agencies. It is also often the case that these organs switch from simple consultation to concerted action and *vice versa*.

In France there is a 'committee of wise men' made up of persons nominated 'because of their qualifications in matters of scientific and technical research and economics'. These persons sit on the committee in their personal capacity and not as representatives of any organization. Concerted action on definite projects, with those responsible for carrying out these projects, is chiefly conducted through planning committees and round tables of the General Delegation for Scientific and Technological Research (DGRST), which include representatives of all interested parties.

In Belgium, the National Science Policy Council, created in 1959, includes persons who are representatives of the scientific, economic and commercial communities. Its statutes laid down that 'Members of the Council sit in their own right and not as agents or delegates of the institutions or organizations to which they belong'. This phrase was deleted in 1966, which marked an evolution from consultation to concerted action. At first the government sought the advice of experts and desired them to be as independent as possible. Today the government rather tries to find at least some guiding principles which would bind the parties, and therefore prefers to confer with fully authorized delegates.

In the United Kingdom the text of the statute which created the new Council for Scientific Policy makes no mention of questions of representation. The same is true of the National Science Policy Council formed in the Netherlands, but it appears from the work which preceded the setting up of this body that the members of the Council sit on it 'in their own right and not as representatives of any institution'.

Each government must choose the formula which suits it best in terms of the advantages that it is looking for and the drawbacks that it wishes to avoid, in the light of its own customs and political traditions. The only consideration that could be added is that the more a government wishes to encourage concerted action binding the partners, the greater is the importance that it ought to attach to the representative character of the persons whom it wishes to enlist and the more numerous ought to be the collective and individual agreements with the partners, who will be fewer in number, but whose commitments will more readily lead to action. The national science plan will appear more like a policy declaration which can later embrace all types of 'development contracts' in science and technology; its formulation will thus concern and bind the interested parties more firmly.

The secretariat necessary for a 'concerted action' policy

So long as it was only a question of consultation, the councils often possessed independent secretariats to study the records and prepare recommendations. During the evolution towards concerted action, this function became progressively more identified with planning in the proper sense of the word. For this reason, when a country evolves from a policy for science to a science-based development policy, it usually happens that the secretariat of the advisory council evolves towards the function of planning in the same degree as the advisory council itself evolves towards a forum for concerted action.[1] The subsequent stage is a closer co-operation or co-ordination of the department of science planning with the over-all national development planning department. However, such a *rapprochement* does not lead to unification, because scientific planning takes place more in the long term than the national development plan. The department of science planning, after it has been co-ordinated with the national development planning department, is then led to assume the function of thinking in the long term about the future of the nation.

SECOND FUNCTION: CO-ORDINATION BETWEEN MINISTRIES

MINISTERIAL FUNCTION

We have mentioned earlier the co-ordination between ministries—a co-ordination which aims at the internal consistency of all parts of the government machine. Though its chief object is connected with the budget, this co-ordination naturally spreads to all aspects of the operation. This makes it necessary that the various departments should accommodate their methods or adjust their objectives to attain such consistency. In matters of science policy, co-ordination between ministries has remained for a long time in an embryonic state.

So long as research was principally carried out in the universities, the Minister of Education found himself at the same time 'Minister of Science',

---

1. Evolution in this direction which has taken place especially in the United States of America, France, Belgium and Canada, shows clearly that this process is necessary.

and the few scientific institutions of the government were often attached to his Ministry. This historical accident gave the Minister of Education of several countries a kind of precedence in scientific matters, even when the other Ministers were, in their turn, invested with greater or lesser scientific responsibilities, which were linked with the operational missions of their department.

But during the last ten or twenty years a distinct ministerial function for science has made its appearance in certain countries. The exercise of this ministerial function has assumed various forms.

The Minister of Science (or of scientific research) can simply be the spokesman for science inside the Cabinet. In this case he exercises hardly any responsibilities of management and has no important budget under his control. His role is personal influence and persuasion. His aim is to promote the general interests of science among his colleagues, who maintain their respective powers in matters of scientific activity. This was the role of the United Kingdom science Minister from 1959 to 1964.[1] He was, in fact, Minister *for* Science.

The Minister of Science (or of scientific research) can in other circumstances unite under his authority the principal research establishments of the government. Thus he assumes direct responsibilities in the management of scientific activities. He would be a fully responsible Minister of Science, and all the government scientific institutions would be answerable to him. But, usually, one or two of the important sectors, such as universities, or technology, or military research, are not. This is the situation which exists in the Federal Republic of Germany.[2] It is also the situation of the Secretary of State for Education and Science in the United Kingdom.

In a certain number of countries the responsibilities of management have remained in the traditional ministries, but inside the government there has been created a ministry to co-ordinate science policy. Since the only one who can co-ordinate is the man who can give orders, it is logical that this function should be assumed by the head of the government, or a minister delegated by him. Such is the case in Belgium[3] and in Sweden, where the Prime Minister himself co-ordinates scientific policy. In France there has existed, since 1969, a Minister of Industrial and Scientific Development (who is also in charge of atomic and space problems and data processing) who holds the presidency of the Interministerial Council for Scientific and Technical Research as the delegate of the Prime Minister. Although he is not

1. In April 1964 the function of the Minister for Science was suppressed. Thereafter two ministers shared 'operational' responsibilities in scientific matters: the Secretary of State (or Minister) for Education and Science was responsible for fundamental research, and the Minister of Technology (now Secretary of State for Trade and Industry) was responsible for research aimed at industrial development.
2. However, the Federal Minister of Research ensures a co-ordination between the scientific activities which are under the jurisdiction of the Länder and those under the Federal Government.
3. In this country, science policy is in the Prime Minister's department. Since 1968 there has been a Minister for the Programming of Scientific Policy, who co-ordinates, in his capacity as delegate of the Prime Minister, all the scientific activities and expenditure of the government.

a minister in the strict sense of the word, the scientific adviser of the President of the United States of America, who is a member of the Executive Bureau of the President, can be compared to a co-ordinating minister.

In the case of a 'Minister of Science' there is a regrouping, so far as possible, of all offices and bureaux which manage the scientific establishments of the government and the research contracts. This deprives the specialized ministerial departments of their scientific staff and their means of research.

Thus agricultural policy is separated from agricultural research, military strategy from armament research, and industrial policy from technological research, etc.

The formula of co-ordination at the level of the head of government keeps the executive and the staff close to the operational offices, but settles the strategic decisions as near the summit as possible.

Controversies sprang up in 1959 in the United States of America about a project to create a Department of Science whose object was to regroup the scientific activities which until then had been under the various departments and federal agencies.[1] The project was later shelved; some argued that to create a new department, one must give it a major objective, and that science could not constitute a major objective;[2] some feared that the Department of Science would in fact exercise control over the activities of other departments, while others dwelt on the drawbacks which would result from separating the activities of research from the other operational activities inside a department with which they were connected.

The two first arguments are clearly contradictory, and the first will hardly be convincing in the case of the United States of America where the scientific budget has become the most important choice that the government and the nation have to make, and a key element in the structure of power. The last argument is the one which will always be valid. If research is a means of action, each ministry and each public department ought to retain 'its own' research: military research ought to belong to the Ministry of Defence, agricultural research to the Ministry of Agriculture, research into road-building to the Ministry of Public Works, educational research to the Ministry of Education, etc.

Thus, the Ministry of Science will only possess a residue of the prerogatives of management, yet this will take away from the head of this department the authority to co-ordinate the use which his colleagues make of their research.

1. There is also an early project of this kind dating from 1880. On the arguments produced for and against a Department of Science, see particularly: Hubert H. Humphrey, 'The Need for a Department of Science', *Annals of the American Academy of Political and Social Science*, January 1960, p. 27-35; Donald W. Cox, *America's New Policy Makers: The Scientists Rise to Power*, p. 157-68. Washington, Chilton Books, 1964.
2. 'Science and technology cannot be said to constitute a major purpose of government', declaration of the Bureau of the Budget, in: *Hearings before the Sub-committee on Reorganization and International Organizations of the Committee on Government Operations, U.S. Senate, May 28, 1959*, part 2, p. 181.

It is always found that a minister who controls institutions, and contracts, and manages operational budgets, loses, by that very fact, his right to oversee similar responsibilities carried by his colleagues. The same is true of the role of arbiter, which the co-ordinator must assume at the time when the national science budget is authorized; at the time of the distribution of the science budget funds he is unable to arbitrate between his own financial requirements and those of other ministers. The acceptance of the prerogatives of management has been shown to be fatal to many ministries of science: in return for having accepted a little empire, made up of some institutes for research and some funds for encouraging research, they have lost their co-ordinating role and now find themselves restricted to their tiny kingdom.

This is, no doubt, part of a profound change in the structure of governments, which is leading to a separation of the functions of the 'staff' from the 'line' functions (or actual managerial operations), a distinction which has always been in use in armies and industrial enterprises. It appears in the form of a new grade of ministers of high political rank (often vice-premiers or ex-prime ministers) who have charge of interministerial co-ordination. These ministers have under them a team of planners or policy-makers. They do not control any operational budget, and this allows them to intervene effectively in budget arbitrations and discussions on the management of appropriations.

Before arriving at this formula, governments often make trial of a hybrid ministry, in charge of both the residual line functions and the staff functions. It is impossible for them to cope with the second. The function of co-ordination becomes unworkable and a proper science policy proves unattainable.

For this reason it appears that ministers of science (and/or technology) are a characteristic of the phase that we have named a 'policy *for* science', while the co-ordinators of science policy belong rather to the phase of development policies based *on* science. The political phase of development based on science could one day make it possible for the same vice-premier to be entrusted with the chairmanship of the ministerial committee for science policy and the ministerial committee for economic policy with the corresponding staffs (the Planning Bureau and the Department for Science Planning), but this should only happen after the countries concerned have first passed through the preceding phases.

MACHINERY FOR THE ADMINISTRATION OF SCIENCE

The existence of an administration for science poses in the main the same problem as that of a Minister of Science.

If the ministerial function is confined to ensuring that science is represented inside the cabinet, separate administrative machinery is unnecessary. This was the case in Italy until 1968, where there was a minister, but no ministry, of science.

In the United Kingdom also, when there was a Minister for Science, he had to be 'kept away from all serious pressure of administrative duties' (Haldane Report).

160

On the other hand if a Minister of Science has direct responsibilities in the carrying out of research, an administrative staff is indispensable for him. His functions consist in administering the establishments, the grants and the contracts; authorizing expenditure; appointing personnel and being responsible for their career prospects; controlling the execution of research, and authorizing a redeployment of resources when it is necessary; concluding contracts; allotting fellowships, subventions, etc. There is a large number of decisions to be made every day, of an operational nature, what are sometimes called 'microdecisions' (in contrast to 'macrodecisions', which have a bearing on the over-all strategy, or on the large sums of money or the portfolios of the budget, or the programmes or plans).

It must be made clear that the management of contracts is a complex and difficult task, which demands a high degree of attainment in scientific or technical knowledge in the subject matter, as well as competence in methods of computerized planning and organization. The value of a government policy depends as much on the quality of its microdecisions as on that of its strategy. No derogatory meaning must therefore be given to the expression 'line'; for the sole aim of this pair of expressions—'staff', and 'line'—is to distinguish the planes on which these functions lie, since the two have been confused with each other for too long, and both have always been inadequately fulfilled.

THE CONTROLLING ORGAN OF SCIENCE POLICY

The administration of research must not be confused with the teams of planners and *policy-makers* which operate under the co-ordinating ministers or under the head of government.

In France, this controlling organ is the Délégation Générale à la Recherche Scientifique et Technique (DGRST), which is placed under the authority of a Delegate General who is answerable to the Minister delegated for research and atomic and space matters. The General Delegation sprang from the transformation of the secretariat of the former Higher Council for Scientific and Technological Research into an independent administrative body. It numbers only about one hundred persons, some thirty of whom are officials with a university education. The latter, who have been carefully chosen, are detached for a specified time by their original administrative body to serve on the General Delegation; this makes it possible to ensure a rotation of personnel and the maintenance of a high level of efficiency.

In the United States of America the staff-function is performed by the Office of Science and Technology, which is part of the Executive Bureau of the President and comprises about fifteen specialists in science and administration. The Office had its origin, as in France, in the secretariat of the consultative organ, the Advisory Council on Science and Technology, which was promoted into an independent administrative body and placed under the direction of the scientific adviser of the President. Like the General Delegation in France, the Office has access to all the files of the departments and agencies. On the other hand the scientific adviser, while being at the same time chairman of the Council (the President's Science Advisory Committee) and of the Federal Council (Federal Council on Science and Technology), has direct access to Congressional Committees, before which he can defend government policy.

In Belgium this function devolves on the Department for the Programming of Science Policy, which is answerable to the Prime Minister. The secretary general of this department is head of the secretariat of the advisory council (National Science Policy Council) and is chairman of the committee of senior civil servants delegated by the various interested ministerial departments (Interministerial Committee for Science Policy).

CO-ORDINATION AND STANDARDIZATION OF THE ADMINISTRATION

In order that the national scientific potential, the administrative machine which directs it, and the budget which finances its growth may together become an effective tool in the hands of the government, they must first be adapted for such an assignment.

In the majority of cases, a preliminary phase of administrative reforms is also necessary. The size of these reforms is restricted, but their difficulty and their extent must not be minimized, for until they have been agreed upon, the macrodecisions will often remain a dead letter, even when they are the policy of the head of the government and have been adopted in the Council of Ministers (or Cabinet).

We have already provided—while dealing with the subject of budgetary analysis—a first example of the administrative standardization which is necessary: one must ensure that every item within a given heading of the science budget contains not only exclusively scientific expenditure, but expenditure of one kind only, so that, when deciding the amount to be entered in respect of this item in the parliamentary budget, the government can effectively regulate the volume of the activity in question.

Another example is that of the essential status of the scientific institutions of the government and their scientific personnel, whatever the ministry on which these institutions depend. Before the standardization of these arrangements is secured, every measure of reorganization and transfer of authority, and every redistribution of power, faces insurmountable difficulties. Further, if the research workers whose activities are financed by contracts or grants do not enjoy a regular and protected status, their insecurity will give rise to a political or social malaise which will in fact oblige the responsible managements to renegotiate the contracts continually and thus make it impossible to manage a dynamic budget. It is therefore a preliminary condition of making the national science budget workable that the personnel should have security of tenure and yet remain highly receptive to change and mobile in their assignments or employment.

A third example is the practice of certain managers of funds for the promotion of science who examine the requests of research workers for support every month of the year, and so give themselves no chance of balancing the support which they give to various branches of science and technology. These managers make it impossible for themselves to see the wood for the trees. Programming can only be introduced into their work by regrouping all the decisions about the granting of loans or subventions on one or two dates every year. For this purpose the records must be prepared longer in advance, and all the agreements must take effect on the same dates. These are small reforms, but without them government finance remains unwieldy and incapable of promoting action. Resistance is sometimes stubborn since everyone clings to the complicated arrangements which protect them from clear government directives.

In practice governments should have the foresight patiently to 'check their gear and their instrument panels' before embarking on an ambitious science plan. They will satisfy themselves that the administrative bodies charged with the duty of managing contracts and government centres are well established and of good quality, that they respond efficiently to the directives which have been given to them at the interministerial level, and that the flow of money is effectively controlled by the mechanism of the programming-budgeting system. Such a task demands several years; like the survey of the national scientific and technological potential, it provides the foundation of all effective science policy.

THIRD FUNCTION: PROMOTION OF RESEARCH AND ITS FINANCING

Promotion is a means to action; for the achievement of the national scientific programme it puts into operation the techniques of financing and supervision.

Some of the methods used by governments to stimulate research are concerned with the general framework in which the research develops; these are the general stimuli. Others consist of particular interventions in favour of a research project or institution, or of an individual research worker, or research team; these are particular stimuli.

The general stimuli include measures concerning patent law, government support for co-operative research, fiscal measures favourable to research, and the organization of scientific and technological information to serve the requirements of research workers.

The particular stimuli include subventions and contracts for research, scientific prizes, and research fellowships.

GENERAL STIMULI

The legal protection of invention

The granting of a temporary monopoly to an inventor by the State has the effect of opening up a gap between him and his competitors, and from the grant he derives an income called royalty. The holder of the patent or those under licence to him, can in fact sell a new invention more dearly since the law disables imitators from selling at a lower price (which the inventor or his

163

licensees would then have to follow). If the patent covers a new process which makes it possible to produce at a lower cost an article which others are already producing, the holder of the patent and his licensees have the liberty of choosing their tactics within the following limits: they may either sell at the others' cost price and still make a profit, or sell nearer their own cost of production, and price the others out of the market. The temporary legal monopoly allows them to choose, within the margin of possible prices defined by these two limits, that which maximizes their profit or their power.

The monopoly is granted by the State in return for the publication of the invention and its effective development, that is for a service rendered by the inventor to the community. The right of holding a patent is thus an exception to the general principle of free competition. The encouragement is granted more at the expense of the consumer than the public treasury. In this respect it is comparable with customs tariffs imposed to encourage industrialization.

However that may be, for many decades, the legal system of capitalist countries and countries with a mixed economy has allowed private individuals and firms to obtain a return on capital invested in original technological innovation. That is the chief and unquestionable merit of the present monopoly system. However, its present evolution poses three problems.

The first concerns the abuses of economic power to which it can give rise; the second, gaps—or rather limitations—in the system; the third, the international consequences of government financing of industrial research.

Abuses of economic power. Monopolies were originally granted for the purpose of providing income. No one then foresaw that they would be used to dominate the market or obtain control of rival enterprises.

Since technological research is no longer the occasional sporadic activity of ingenious or gifted individuals—since it has become a constant activity of the large laboratories of industrial firms, which are organized for technological innovation in the same way that other departments of the same firms are organized for producing and selling—the patent has become a chosen weapon in struggles for power between firms.

None of the traditional sectors of industry has produced enterprises of the vast size that we have come to know in the two chief industries based on science, namely chemistry and electrotechnics. Although it is true that savings from large-scale operation have thrown up firms of from 5,000 to 20,000 workers in the metallurgical industries, the leading firms of newer industries can nowadays often group ten times as many workers. In these new branches of industry, the patents portfolio and the research laboratory have often become the decisive weapons which put other competitor firms out of business. The trouble is not that the production of these smaller firms is insufficient in volume for them to remain economically viable; it is simply that they cannot compete in respect of patents and research. Yet one hesitates to speak of the abuse of economic power when the subject is a monopoly created by law, although some people think that this monopoly ought to be used to give a return for investment in research and not to build empires.

Their thoughts turn towards compulsory licensing, a rule which is applied in certain countries to the holders of foreign patents. The application of this principle has however remained restricted, probably because it can be as easily used for the profit of the strongest as for the advantage of the weakest; in fact a business of medium size could be deprived of the special advantage that it has in exploiting and marketing the article whose patent it has taken out: a powerful firm could get itself issued with a compulsory licence and enter the market before the firm which holds the patent. Indeed, the means of production and sale of the large firms are so superior that they could easily cut the ground from under the feet of their small inventive competitors, even while paying them dues. This is probably why the laws governing the protection of invention have not been revised in principle despite the almost complete transformation of the industrial setting to which they apply, and despite the obvious uneasiness about the use to which these laws are put— namely the acquisition of positions of industrial domination.

Limitations. For reasons which are fairly evident, protection is not granted in respect of the discovery of general scientific principles, but only in respect of particular applications thereof. The result is, however, that scientific breakthroughs, in the full sense of the word, can rarely be protected, at least in the realm of physics. This can be done more easily in chemistry, because the process by which a new compound is obtained is capable of being patented, and the number of possible ways of forming the compound is often limited. With this as the only big exception, patents give more protection to expenditure on experimental development than they do to expenditure on research.

Other investments in research and experimental development are also left without proper protection, for example, those where a firm could not easily collect its patent dues. This is the case in agricultural research. Long-term developments, such as the reactor families for power stations, constitute another such example.

It can be seen from these examples that patent law is far from ensuring a return for all expenditure on technological research which is necessary for the progress of the economy.

The industrial world has met this situation in two ways: first, by asking the government to take charge of the expenditure in question; secondly, by organizing firms to exploit quickly—before competitors have time to react— the necessarily short-lived industrial and commercial advantage of discoveries that cannot be patented. Royalties are thus replaced by 'profit from technological innovation', which lasts an appreciably shorter time, but can be constantly reproduced by an appropriate business strategy. The same procedure is recommended when the invention is capable of being patented, but is in danger of losing its commercial value because of parallel breakthroughs achieved elsewhere.

In this strategy, the profits or the power no longer proceed principally from the patents which are exploited, but from the obstacles put in the way of competitors to cause them delays and reverses: this is why the defensive

165

patent has become as important as the invention patent proper—sometimes indeed more so.

There is evident cause for alarm in a practice which has so considerably perverted the original institution and the moral and legal foundations on which it rests. However, patent law is part of an economic system of 'Schumpeterian' competition which ensures, such as it is, a rapid flow of innovation. This—very pragmatic—consideration tends to discourage attempts at radical reform, or indeed the abolition of the legal monopoly. A more empirical struggle against the abuses of power appears to be preferable.

International consequences of government financing of industrial research. International trade continues, in a large part of the world, to be based on the free play of competition. Patent law, as has just been seen, is an exception to this—an exception whose dimensions have already been increased by the unequal distribution of the research potential of the different countries. But government financing of industrial research presents a second exception more important still than the first, since the financial neutrality of government in the competitive areas of the economy is a golden rule of free-market economics.

Whatever the reasons which may lead a government to take charge of industrial research—and the present work points out more than once that there are excellent reasons for its doing so—it cannot be denied that by so acting a government creates a bias in the conditions of international competition. Other countries, whose industries have to suffer from this bias, would be theoretically entitled to defend themselves by protectionist measures, if they have not the resources to re-establish complete equality of opportunity by the same weapons—that is, by means of compensation paid to their own industries in the form of subventions to their scientific research.

It would clearly be more reasonable to ask for compensation from the foreign governments whose action has caused the bias in competition; and it would no doubt be appropriate to do this in the name of the principles which those same governments often profess. The best compensation—and the one that would be the easiest to grant by international agreement— would be to grant generous rights of access to the results of government-financed research. Countries could give each other mutual assurance of reciprocal rights of access. And since the balance of advantages which each would draw from the arrangement would probably not be equal, the most advanced countries would also discover, in such arrangements, opportunity for giving very effective aid to all the others. Indeed, the provision of such aid would seem to be no more than equity demands.

The application of this principle would be relatively easy. Whenever a firm in country A—let us call it the 'applicant firm'—applied for a patent in country B, an investigation would be made with a view to determining what government contracts and government subsidies the applicant firm (or the firm from which it holds rights) has enjoyed in country A, during the period preceding the first application for a patent. If it appeared that the patent was in some degree the result of government contracts or subsidies in country A, then the applicant firm would be invited to put licences publicly on offer, on fair and reasonable terms, in country B. This offer would then entitle the applicant firm to register the patent in country B.

Government aid to co-operative research. In several countries, industrial enterprises—especially those which do not possess enough resources to pursue effective research on their own account—have banded together to form collective organizations for research. Because of the help which these co-operative methods can provide in developing all the research activity in one industrial sector, certain governments have passed laws to allow the public authorities to extend this formula: sometimes by establishing new centres themselves, and sometimes by encouraging their establishment by granting them a special status, which allows tax concessions.[1]

Government action thus aims at ensuring them regular financial help, while preserving the principle of community management of a co-operative kind. The extent of a government's participation in this field varies greatly from one country to another. In some countries, such as France, the State has virtually confined itself to granting these centres an official status. In others, generous financial aid has been granted as well as this protection. Sometimes, a government has allowed them to levy a 'compulsory quota', which amounts to an outright tax on the firms of a particular industrial sector.

Until its assignment in this field was taken over by the Ministry of Technology (now, Department of Trade and Industry), the Department of Scientific and Industrial Research had in fact established in the United Kingdom, even before the Second World War, a large network of research associations.

In France, a law of 1948 allows the government to set up Technical Industrial Centres in every branch of activity where the general interest requires it. Such a step is taken only after advice[2] has been received from the most representative of the employers' federations, of professional associations and trade unions in the field concerned. When giving its approval for the setting up of the proposed centre, the executive power defines the limits of the economic sector or industrial branch in question.

The financial support which each government gives to these centres can assume various forms.

In France, the government has imposed payment of a compulsory quota by all enterprises in that branch of industry. These quotas, established by statute, are recognized as belonging legally to the system of financial law, and are considered as 'parafiscal taxes'. These resources may be added to by grants and loans given by the government.

1. On this subject see: F. N. Woodward, *Structure of Industrial Research Associations,* and, *Industrial Research Associations in France, Belgium and Germany,* Paris, OECD, 1965.
2. The original draft envisaged 'the agreement' of these organizations, but the government's power of decision has been strengthened by changing this to a simple 'opinion' —which the government is not obliged to follow.

In Belgium a similar system is in use: a royal decree determines the annual contribution which must be paid by all firms in the 'combine', in proportion to their importance. Thus, in these two countries the government directly orders that all the firms in one economic sector or industrial branch shall affiliate themselves to the research centres.[1] But, whereas in France the direct financial support provided by the government remains of small importance compared with the private contributions, in Belgium the government subsidies granted to the co-operative research centres represent about one-third of the resources of these centres.

Conversely, in the United Kingdom, a system of 'parafiscality' inaugurated in 1945—that is, at the same time as in Belgium and France—has not been successful and has had to be replaced by a system of voluntary adherence.

The British Government gives the research associations a block grant for five years, and, according to the circumstances, a supplementary grant[2] to allow the associations to extend their activities, but the whole system reveals a great flexibility, since by anticipating a reduction in government support, it allows a progressive takeover of responsibility for financial contribution to the association by the member companies.

It is important however to see what is the actual effect of government intervention in this field. In considering this, one assumes that co-operative research aims especially at improving present technology and products, and rarely at inventing new manufacturing processes and new products. The reason for this is simple. The programmes of the co-operative research centres are designed to be adapted to the need of a number of firms, which is often very large. The most usual kind of study will then be research on raw materials and the improvement of manufacturing processes, because this interests all the firms of an industrial sector to the same extent.

As J. M. Collette has remarked, 'co-operative research is not therefore an activity which produces the main discoveries and inventions that cause a

1. F. N. Woodward points out nevertheless that 'there has been a great deal of criticism of the system of compulsory taxation, because, with this system, the members' enthusiasm fluctuates and the direction of the research association, possessed of an assured income, is thus made less dynamic' (*Industrial Research Associations in France, Belgium and Germany*, op. cit., p. 26).
2. Originally, subventions were all appropriated for the general aims of the association and not for special research tasks in which the government might have been interested. The total of a block grant was generally calculated with regard to the sum which the associated firms were themselves ready to contribute. A recent practice, which has just been adopted and may perhaps be extended, consists in granting additional 'special' subventions, with the aim of encouraging particular pieces of research which appear to be in the national interest. This new technique is especially interesting, as it indicates the general evolution of a way of thinking which can spread to other organizations like the Research Councils; the main result would be that instead of granting subventions in response to the requests of scientists, the government would take the initiative in promoting special fields and projects of research by the offer of subventions. (*National Science Policies: United Kingdom, Germany*, Paris, OECD, 1966.)

spectacular advance in existing technology or which transform the shape of a market by bringing out new products'.[1]

In essence, co-operative research is a service to industrial firms. This characteristic explains the great differences that can be observed in the part which co-operative research plays in different sectors.

An OECD report emphasizes that 'the activities of co-operative research institutes have only small importance in the science-based industries, such as chemical products and electrical construction and scientific instruments. On the other hand they assume a great importance in the traditional industries like textiles, foodstuffs, glass, cement, pottery, wood and paper ... metallurgy, oil ... and naval construction'.

Co-operative research appears to be a stage in the support granted by the government to industrial research. It is essential during the first phase of industrialization.

In the countries which are starting or are engaged in the second phase of industrialization, which is characterized by technological innovation, governments endeavour increasingly—by a different method, the method of contracts—to finance the 'competitive' research for which secrecy is an essential condition of success. But this new orientation of government action entails procedures that are considerably more complex, and brings to the forefront much more important principles and concepts. One might say that on this level the majority of governments are still groping in the dark.

Promoting of research by fiscal means

It is generally agreed that tax deductions are an important means of encouraging research activity.

Some people even show a distinct preference for this formula, which has, in their eyes, the merit of preserving the free choice of those concerned, compared with the contract procedure which is by definition selective and which they consider discriminatory. Others think that the contract or subvention confers more direct advantages, and makes it possible to respond to more particular needs, while at the same time being less costly to the Treasury than a general reduction in taxation.

This argument applies, it is true, to all the activities which a government may wish to encourage or develop. Relief from taxation usually causes an appreciably greater loss of revenue than the extra cost involved in directly financing the desired activity. But of all stimuli, relief from taxation is the one which gives the technostructure of the enterprise most freedom from bureaucratic interference by a government.

In saying this, one takes it for granted that hitherto few countries have employed fiscal measures to promote research which would flatly contradict the precepts of common law. On this subject a government adopts an attitude which is little more than a 'liberal' interpretation of legal limits. A Treasury does not easily rid itself of the fear that fiscal advantages granted to research

---

1. J. M. Collette, 'La Recherche, Développement en Grande-Bretagne', in: *Cahiers de l'Institut des Sciences Economiques Appliquées,* Paris, series T, no. 2, p. 138.

may actually cover avoidance of tax by other activities. Fiscal law concerning research is therefore a collection of empirical rules, and nowhere is there to be found a 'fiscal code for research'.

The complexity and diversity of practice in this field enhance the interest of the synthesis which has been attempted by J. Van Hoorn, in a report which he drew up for OECD in 1961.[1] Although in this subject things change quickly, the very comprehensive picture that he has painted makes it possible for one to get a clear idea of the chief problems raised by research in the field of tax assessment. We have made extensive use of this work in the following rapid survey of the fiscal problems of research.

Taxes on the income of individuals and on company profits. As regards running expenses assignable to industrial research (salaries, wages, purchase of articles necessary for research) it is assumed that these may be deducted from taxable income under the conditions which usually apply to the calculation of that income. The same is usually true of the sums paid by industrial enterprises in financing contracts for research, provided that such research is carried out in the interest of the enterprise and is not an independent effort.

As regards expenses which are of the character of capital expenditure, the usually accepted arrangement in a nation's law is that they are not deductible in the claim for the financial year in the course of which they become liable to tax, but annual deductions are authorized for amortization; besides this, in some countries the incidence of the amortization can be speeded up.

The usual tables of amortization are not suitable for research: the apparatus used in research as a general rule becomes rapidly out of date, especially in the case of pilot plant and 'full-scale' prototypes. The United Kingdom, Canada and Japan have applied in this sphere an unusually favourable formula: the buildings, plant and material allotted for research can be amortized at a higher price than the actual cost price. This system of supplementary deductions becomes in fact a subsidy—amounting to 130 per cent in Canada, 150 per cent in the United Kingdom and 195 per cent in Japan.

As regards gifts made to independent research institutions (which have no connexion with the donor) most countries lay down that these sums can be deducted from the profits of the current financial year, though a ceiling is sometimes fixed.

The problem of research centres working under contract is an important one: should the excess of their receipts over their expenditure be counted as a working profit and taxed as such?

Taxes on turnover and added value. J. Van Hoorn points out that in certain countries services are not liable to the tax on turnover, so that the services

1. J. Van Hoorn, *Régime Fiscal de la Recherche et du Développement Technique*, Paris, OECD, 1961.

rendered by research laboratories or other similar institutions are free of tax. The transfer of plans, models, drawings or any rights, etc., by the inventor or maker are generally considered as a sale of services. In countries where services are not exempt from tax on turnover, it is sometimes considered that research services rendered by scientific institutions have been conducted in the public interest, and these activities have therefore been exempted from tax.[1]

The delivery of products and the sales of equipment to scientific centres for research purposes are exempt from taxes in some countries but not in others.[2] However, for simplicity's sake it is the tendency in several countries to apply the tax on consumption to all sales, even if the buyer is the government. Exceptions made in favour of other privileged buyers because of the disinterested nature of their activities disappear in this case.

Regulations which apply to shareholding in research companies. Certain countries (for example, France) have provided for an exceptionally large amortization of shares or interest acquired at the time of the subscription of the initial capital or the further raisings of capital for research companies or organizations. These regulations ought to make it possible to ensure equality of treatment in taxation to the two choices which usually face large enterprises: either to build and equip a new laboratory on their own premises, or to transfer the capital to a research company.

Taxation on imported research equipment and instruments. In many countries, instruments and equipment intended for research can be imported free of custom duties. However, in most countries, this exemption is applicable to non-profit-making organizations only.

Taxation on the product of research: regulations applicable to sale of patents and granting of licences. There is a strong tendency to regard the transfer of intangible rights as a payment for services and taxable as such. The receipt of dues will also be considered as income from personal property. All the conditions for cumulative taxation are therefore brought to bear on the research enterprise, that is the enterprise which does not amortize the expenses of its laboratories against the profits derived from the production of goods, but chooses rather to put a value on its inventions by the sale of patents, licences and 'know-how'. This is one of the most complex of subjects and one most unsatisfactorily dealt with by fiscal legislation.

THE MOST IMPORTANT OF THE SPECIAL STIMULI: SUBVENTIONS AND CONTRACTS

The distinction between the subvention for research and the contract for research and experimental development

The subvention is an allocation granted for general purposes so that the scientific results obtained by the recipient do not—as in the case of the

---

1. J. Van Hoorn, op. cit., p. 27.
2. ibid., p. 28.

contract for research—constitute an actual counterpart for the sums granted. It is a kind of donation to further the achievement of research projects or to pay for research equipment.

The subvention gives the recipient great freedom. More often than not, his only obligations are to submit a report on the research activity and assessment of its results, interim assessments during the course of it, and a final assessment on its completion. The subvention is in fact a unilateral commitment on the part of the financing organization within the framework of conditions determined in advance before the grant is made. It is the usual method of supporting fundamental research, particularly university research. It is also used to help co-operative research with industrial aims. It may be allocated either to an individual or to a team for the achievement of a special project. It may also be allocated to a department, or even an institution for the whole of its research programme. In this last case it is called a block grant. It may hold good for one or several years, and may cover the whole expense or only part of it.

In several countries a large network of contracts for research has been created, especially since the Second World War; these are concluded by the central government with private firms or institutions, or even, as we shall see later, with government organizations. In the United States of America, this system is the basis of the governmental science policy, since the essential activity of Federal Agencies in research matters is the arrangement of the contracts which they enter into with outside organizations. Federal research contracts involved, in 1966, an expenditure of 5,500 million dollars.

At first, governments had recourse to this procedure in the role of buyer in response to the particular needs of administrative bodies, which were usually for modern equipment: weapons, aeroplanes, rockets, computers, etc. Lately, research contracts have been passed by governments for general promotion of research, and no longer in answer to the special needs of certain administrative bodies. Their object is now to encourage research activity in the nation in terms of national objectives.[1] In the first case, one is dealing with research contracts linked to government purchases of equipment and manufactured articles; in the second case, the object of the contracts is to encourage progress, an end which is akin to that of subventions.

Even if, in actual fact, the borderline between subvention and contract is sometimes blurred, there does nevertheless exist, between these two types of financing, a difference in nature.

1. See in: *Aspects Juridiques de la Recherche Scientifique* (Faculty of Law of Liège, and Martinus Nijhoff, The Hague, 1965), 'Les marchés d'étude et de recherche passés par les administrations publiques' (French regulations, described by P. Huvelin, p. 149-65, and regulations of the United States of America described by G. Pevtchin, p. 167-217).

See also: 'Les Modalités d'Aide de l'État aux Entreprises du Secteur Concurrentiel en Matière de Recherche Scientifique et Technique', *Le Progrès Scientifique*, no. 97, June 1966. Paris, DGRST. A comprehensive bibliography on research contracts in the United States of America, collected together by G. Pevtchin, appears on pages 411-16 of *Aspects Juridiques de la Recherche Scientifique*.

In the case of the subvention, the initiative belongs to those performing the research. Anyone, in principle—provided that he agrees to conform to the general conditions laid down—can ask for a subvention, provided he has the capacity to undertake the project which he is putting forward.

In the case of a research contract the initiative is, strictly speaking, that of the sponsor, i.e. the State, even if it falls in with the suggestions of those performing the research. By its very nature, contract research is therefore directed research. As his side of the bargain, the research worker provides the results, the conclusions, the knowledge, and sometimes also even proprietary rights in the fruits of his research.

The choice of the contracting party corresponds with the wish of the sponsor to see the best possible work carried out, in the shortest time and at the best price.

Classification of contracts according to their ultimate objective

Research contracts linked to purchase of supplies of 'software'. This is a new policy adopted by governments. It has assumed considerable importance in certain countries under the impulse of military needs. Its aim is to make the financing of the 'software' (studies and prototype work) separate from that of the 'hardware' (mass production and delivery) in a government order. Contracting for such 'software' is, in fact, applicable for the preliminary studies which can lead directly to industrial production or to construction whenever there is uncertainty about the success of new manufactures, or the construction of new 'hardware', especially when the administration itself is not prepared to undertake the research needed to dispel these uncertainties.[1] The growth of contract research stems from governments finding it impossible to have all the planning and research necessary carried out in their own departments, laboratories or arsenals.

The spectacular development of research contracts in the United States of America appears to be due to the fact that the Federal Government did not create, before the Second World War, a network of government laboratories of the kind that existed in France and in the United Kingdom. The United States Government was therefore compelled to make appropriate arrangements with independent firms and organizations in order to entrust them with the responsibility of carrying out the large research programmes which had become an indispensable part of national policy.

Later it became clear that the choice of industry as the research-performing agent was a happy one. The technological breakthroughs obtained in industrial enterprises rapidly extended to the civilian field, so that the economy received a more decisive advance from the national technological programmes than if these had been restricted to government arsenals.

Contracts for the promotion of research. The need for such contracts can appear either at the level of fundamental research or at that of technological research.

1. See *Le Progrès Scientifique*, no. 97, June 1966, p. 8. Paris, DGRST.

In the first case, their object is to initiate research in new scientific fields whose exploration requires resources—especially in terms of capital equipment—which are far beyond the regular budgets that are available to existing research teams.

In the second case, their object is to help industrial enterprises which show an inclination to make a special effort in the scientific field, if, through lack of resources, they cannot hope to succeed alone; or they may direct industry towards subjects whose commercial pay-off still appears uncertain or too distant for industrial enterprises to take the risk. Nevertheless, if the research is successful, such contracts will open up new channels of development for the national economy. The objective is not only the product itself, but the economic power and industrial vitality which results from the technological breakthrough.

Initiatives of this kind are taken by certain governments with the general aim of supplementing the co-operative forms of scientific activity (research associations) in the more recent sectors of industrial activity (such as machine tools, electrical equipment and appliances, and electronics) where it is the originality of the product (and from this arises the secrecy surrounding its preparation) which is the determining factor of success in industrial competition. Competitive research, which does not lend itself well to co-operative schemes, can be carried out under special contracts negotiated between the sponsoring public authority and the interested industrial enterprises.

But this kind of contract usually raises great difficulties. It is no longer an agreement to pay for the requirements of an administrative body, which is the customer. The contracts are for pioneering research in new fields where the risks of failure are greatest and the prospects of commercial exploitation the most remote. These contracts are therefore utilized to launch research activities in new directions. The government is anticipating the future needs of industry: it must be able to detect the fields in which government intervention is indispensable, in order to select those research projects whose results will help to foster its economic policy.

Such contractual procedures require both scientific and technical knowledge, as well as economic insight on the part of government.

Furthermore, for contracts negotiated with industry, it is important to settle the amount of the government's financial contribution. The formula usually adopted is for an industrial enterprise to cover 50 per cent of the expenditure; this rule does not always give satisfaction, since a half share still represents too heavy a load for projects whose profits are either highly questionable or only to be expected in the very distant future. In certain countries, government financing can reach a much greater proportion of the research cost. However, contracts for the promotion of research should not, as a rule, foresee full cost coverage by the government. In a competitive economy it seems advisable to retain the stimulus of risk.

Another problem is the way in which a government will be repaid in the event of successful research leading to profitable commercial exploitation.

It is generally accepted that in the case of success, the government should receive some return because of the risk that it has run. This can assume various forms: the taking out of patents in the name of the government; an obligation binding the contracting firm to issue licences to a third party of the same nation (with or without payment); repayment with or without interest of the sums committed by the government (possibly under some system similar to licence royalties proportional to sales); reserving to the government a royalty-free licensing right to cover its own needs; government participation in the profits that may accrue from exploiting research results, etc.

In France, there are two categories of contracts for the promotion of research. 'Contrats de recherche' are allotted either by the Prime Minister or by the Defence Minister. The Prime Minister's contracts are administered by the Délégation Générale à la Recherche Scientifique et Technique within the framework of a definite number of research themes which are of national importance and are determined by the national development plan ('actions concertées'). The Defence Minister's contracts are administered by the Division des Recherches et des Moyens d'Essai (DRME) within the framework of certain guiding themes ('thèmes orienteurs') which play the same role as the concerted actions. These research contracts can cover both fundamental research and applied research; they are negotiated both with private enterprises and with government research organizations.

In the United Kingdom it was the Ministry of Technology (now, Department of Trade and Industry) which awarded private enterprises contracts for development in the civilian field, more especially in the industrial sectors over which it exercised 'sponsorship' (a kind of trusteeship): mechanical and electrical manufacture, electronics, telecommunications, etc.

In France, as well as in the United Kingdom, the system of contracts for the promotion of research is of quite recent date. It is as yet difficult to assess the actual impact of this system on the development of research, or to foresee how far its results will tally with the calculations of its promoters.

In the United States of America, attempts have been made by the administration to establish direct Federal aid for technological innovation in industrial sectors which have so far been outside the orbit of government contracts for defence and space, and which consequently suffer from comparative stagnation in technical matters.[1]

The very existence of these plans represents a turning-point in the science policy of the United States. The activating role played by government contracts in industrial development was learnt empirically through 'spin-off' from military, nuclear and space contracts. The government has thus been prompted to request the necessary authorizations and financial resources to spread this kind of promotional action to all sectors of the national life where the need makes itself felt.

Implications of research contracts

The short period since the introduction of government research contracts makes evaluation of this procedure difficult, in terms of its impact on the economy of a country.

---

1. *Science,* 24 September 1965, p. 1485.

The example of the United States is outstanding when one wishes to unravel the effects of the contractual system not only on the proper functions and role of government bodies and contracting organizations (universities and private enterprises), but also on the kind of relationship which emerges between these government bodies and the contractors. The course of events in the United States will accordingly provide us with interesting examples.[1]

It is true that the effects described below are mainly due to peculiarities of the political and economic system of the United States; in this sense they represent a unique situation. One cannot help thinking, however, that in many ways the experience of the United States foreshadows the situation which could prevail in other countries as a result of a large and sharply focused contractual promotion of research by the public authorities.

Implications for the public authority. By means of contracts, government departments largely delegate their responsibilities in technical and scientific matters to independent institutions.

The experience of various countries indicates that it is not necessary for all the civil servants engaged in administration of research contracts to have a scientific education. But it is necessary that they should be able to grasp the technical implications of these contracts, and that they should also have the proper administrative experience. In the words of an official United States report: 'There must exist within the administration sufficient technical knowledge to prevent the technical adviser who comes from outside becoming *de facto* the one who takes the decisions.'[2]

In this respect, the United States administration appears to have an important advantage: the position of civil servants there is less bound by tradition than in other countries. It is a frequent custom in the United States to nominate for high positions in the civil service prominent persons from the private sector or from universities, and this makes it easy to call on persons well qualified in scientific matters when these are not available within the framework of the administration.[3]

The system of contracts also entails important legal consequences with regard to regulation of the government's commercial dealings.

Since the outcome of a piece of research is always uncertain at the start, the output demanded by a government cannot be defined precisely in advance, especially in the field of fundamental research. In the words of

1. On this subject see, especially, the following works: John K. Galbraith, *The New Industrial State*, Boston, Mass., Houghton Mifflin Co., 1967; *Report of the Select Committee on Government Research of the House of Representatives, 88th session*, Washington, United States Government Printing Office, 1964; *Impact of Federal Research and Development Program, Report to the President on Government Contracting for Research and Development*, Washington, Bureau of the Budget, 30 April 1962; Carl F. Stover, *The Government Contract System as a Problem in Public Policy. The Industry-Government Aerospace Relationship*, Stanford Research Institute, January 1963.
2. *Report to the President on Government contracting for Research and Development*, op. cit., p. 20.
3. Don K. Price, *Government and Science*, p. 22-7. New York University Press, 1954.

L. Favoreu, 'this is not an obstacle to the possibility of a conventional commitment, because a contract does not only give rise to *obligations to achieve results;* it can also produce mere *obligations to exert one's abilities,* which are also called *obligations of effort;* these demand on the part of the contractor exertions proportionate to the circumstances and to his potential capacities; they do not oblige him to get a particular result, but to do something or act in a particular way'.[1] But covenants of this kind are as yet hardly seen in government agreements. Consequently the research contract does not quite come into the traditional classifications of the government contract.

Even so, it seems difficult to make the idea of a research contract fit into the usual idea of a market deal. Indeed, the research which is the object of the contract requires a combination of human and material factors such as can only be found in a very small number of industrial enterprises. The personality of the contractor becomes all-important, and considerations of price are put into the background; this is contrary to accepted administrative procedure where the terms of reference are conventional enough to allow a large number of industrial enterprises to compete against each other.[2]

For these reasons there is a tendency today for administrations in this field to turn away from the practice of auctioning contracts, in favour of direct contractual arrangements. In the United States, none of the research contracts authorized by the government are put up for auction.[3] In France, a new regulation for the government's market dealings, which was adopted in 1962 and 1963, provided for an appeal for tenders 'as far as was reasonably possible'. In the Federal Republic of Germany, a special procedure was also provided for.

Several international research organizations have attempted to discover a fair method of placing contracts for research and experimental development based on an appeal for tenders, classifying the tenders received from the lowest price upwards, as is the custom in the government auctions of contracts. These attempts proved fruitless since, in the final reckoning, the price paid to the contractor usually bore no relation to the price originally stated by him in his 'tender'.

Another case is when the administrative body negotiates the agreement with an organization which is under government control. This is what happens in France in the case of the 'actions concertées' administered by the General Delegation which arranges contracts—up to now almost entirely in the sphere of fundamental research—with university departments and even with government institutions for research. However, the legal aspect of the

1. L. Favoreu, 'Un Contrat Administratif de Type Nouveau? Les Conventions de Recherche de la DGRST et de la DRME', in: *Actualité Juridique, Droit Administratif,* 20 September 1965, p. 448.
2. Thus, in the field of space research, the problem is to obtain the maximum reliability at any price, that is the greatest possible degree of infallibility in the working of the equipment ordered from the industry.
3. This reform was initiated in the United States by a law passed in 1941 which released administrative bodies from accepted contractual procedures in light of the needs of war. Afterwards, this decision was made permanent.

dealings between government bodies does not fall within the scope of agreements connected with research.

The new technique for research contracts also raised problems of the rights and obligations of the parties with regard to industrial ownership, when an invention resulting from the research can be patented. The core of the problem is to decide who is to possess the rights over the inventions and discoveries that result from research work financed by a government: is it to be the government or the contractor; and if it is to be the contractor, what conditions are to be attached?

One school of thought maintains that the contractor should keep the right to hold the patent in his own name on the condition that he releases a free licence to the government to exploit this invention.

The following are the main arguments supporting this thesis: (a) the government has no need for a patent because it has neither the right nor the power to use the results of the research itself for commercial ends; (b) research done on a government's behalf is not aimed at commercial adaptation, hence additional development work will be necessary before the invention can be put on the market. Such additional developmental research may be long and costly; under these conditions, unless the industrial enterprise is given the assurance that it will have exclusive rights in the invention, it will not risk involvement in this phase of development; (c) the very fact of knowing that it will possess the right to take out the patent will provide the industrial enterprise with an incentive to put forward its tender.

The opposite school of thought argues as follows: (a) that the research has been paid for with the taxpayer's money and therefore the nation has the right of ownership over all the fruit of this research, including inventions that can be patented; the opposite course would be equivalent to making the taxpayer pay twice, first when he pays for the research, and secondly when he buys the product; (b) if the government possesses the right to the patents, it can spread the results of the research over a wide area, and thus provide a profit for all the industrial enterprises interested in discovery or invention; this will allow the government to give them compensation, in case they feel discriminated against in the matter of the contract.

The United States Government has not yet taken up a definite position in this matter. A regulation of the Kennedy administration, in its anxiety to bring order into a situation which had been described as 'chaotic',[1] established a system that was certainly less complicated but still inconsistent. Its effect was that certain Federal Agencies applied the principles of the first thesis, while others applied those of the second thesis. A summary of this regulation is given as follows by the review *Chemical and Engineering News:*[2] 'In most cases the contractor will take title to an invention resulting from a government R & D contract; the Government will get a royalty-free license to use the invention. In these cases, the contractor must develop the invention or lose his exclusive rights. If, within three years after the patent issues, a contractor has not taken effective steps to bring the invention to the point of

1. Samuel W. Bryant, 'The Patent Mess', *Fortune,* September 1962.
2. 15 August 1966, p. 36.

practical application or made licenses available to others on reasonable terms, the Government will have the right to force the contractor to grant a license to an applicant on a non-exclusive royalty-free basis.

'The Government will take title to patents where:

— The contract is in a field in which the Government has done most of the work or financed most of the work.

— The contract explores fields which directly concern public health or welfare.

— The contract is to develop commercial items which government regulations may require the public to use.

— The services of the contractor are for operating a government-owned research or production facility or for co-ordinating or directing the work of others.'

In the Federal Republic of Germany, when the contract is one for the Ministry of Defence, the contracting firm must grant a licence to the government, but may charge a price. The contracts negotiated by the Federal Ministry of Research make it obligatory for the contractor to grant a right of free licence to the State for all purposes that it considers expedient for its needs and for the progress of science and technology; these licences the government can grant against payment to firms which desire the right to make use of these patents. (Ministry's new name: Education and Science.)

In France, the rights of industrial ownership are, in principle, reserved for the contracting firms, but the State may reserve to itself the whole or part of these rights, by an explicit clause inserted into the agreement: this is a fairly flexible system which underlines the importance of the specifications.

As has been seen, contracts for research and experimental development entail a sweeping change in legal situations in the sphere of government agreements. Few governments have as yet made the effort to introduce the changes and clarifications which this branch of law clearly requires.

Implications of the contracts for recipient institutions. The policy of government contracts tend to give an advantage to certain industrial sectors; for example, different kinds of processing of metals (chiefly aeronautical construction, electrical appliances, electronics, precision mechanics) and less frequently certain branches of chemistry and metallurgy (synthetic products and non-ferrous or rare metals). In these industries, government finance in the United States covers a large part, in some cases virtually the whole, of the expenses of research. These are the industries which contribute most to the production of military equipment and the construction of spacecraft. It is in these industries also that measures taken to promote research seem to have the most useful effect, because of the part which science plays in technological innovation, which is a necessary element of their commercial strategy. It is on these industries that the governments concentrate contracts for promotion of research in France and the United Kingdom. It is also an established fact that in the United States, as well as in France and the United Kingdom, government contracts give an advantage to the larger enterprises.

Both in the military sphere and in the space programme, administrative bodies find themselves very often under the necessity of ordering technological systems, that is simultaneous and integrated development of various elements: a bomber, for example, which cannot be designed without consideration being taken of its load of equipment, weapons and ammunition, or a manned multistage rocket which carries a whole load of precision instruments. The administrative bodies are therefore compelled to grant global contracts to firms that are able to combine the development of many constituents as well as the integrated whole. These leading contractors keep for themselves certain parts of the work which is to be done, and the rest is subcontracted to smaller firms. As a rule, only the large industrial groups are able to assume the responsibilities of leading contractor.

A contract for research also gives the contracting company an overwhelming advantage: in acquiring the 'know-how', a company places itself in a privileged position to claim a contract for production.

A third consequence of government policy in the United States must equally be mentioned: hitherto, research contracts have gone to industries situated in certain areas (California, the south-east and the areas around Boston and Philadelphia) and therefore have tended to increase the existing imbalances in the geographical distribution of the country's industry.

As regards the part played by the universities, we shall also refer to the situation in the United States, since the Federal Government has made considerable use of university laboratories in order to carry out its scientific programmes in the field of nuclear and space research.

It often happens in the United States that the whole of the energies of a university laboratory are absorbed in carrying out a Federal programme. This clearly involves a danger of throwing out of balance the scientific activities of the university in question. It is indeed a strong temptation for the professors and their assistants to abandon academic activities in favour of carrying out a Federal contract which usually involves a considerable amount of development work; this is more remunerative, and ensures research facilities which they would have difficulty in obtaining through academic channels. This situation has often been criticized in the United States, and several official reports have underlined the need to preserve better conditions for fundamental research in the institutions of higher education. The action of the National Science Foundation has undoubtedly made possible an increase in the resources allotted to fundamental research, but the granting of technological contracts to universities has also been on the rise, so that the relative share of the Federal grants allotted to fundamental research has not appreciably increased in the last few years.

In the university sector one can also observe a phenomenon similar to what is happening in the industrial sector; Federal aid is concentrated on a few large universities, situated in certain regions. The fact that these universities possess the best professors makes it easier for them to attract government funds. The reputation and the resources which government favour entails, bring them in turn the best research workers.

Both in industry and in the universities, therefore, government contracts have undoubtedly played an activating role; the large industrial firms, the pioneer sectors, the most fully developed regions, the most powerful universities have all made exceptional progress. This speeding-up process seems to have increased some inequalities, but one must also take into account the total impetus given to the national economy and the indirect advantages which the less favoured sectors have received.

Influence of government research contracts on the national economy

It has been seen that, as things are at present, government research contracts in the United States aim at non-economic objectives; they are mostly connected with the military, atomic and space programmes, though a few have social, cultural and humanitarian aims.

The question arises whether, after all, this accumulation of resources in support of such programmes, and their concentration on certain sectors of industrial activity, does have a beneficial effect on the economic and social plane.

Does all the effort expended in ensuring a lead in the nuclear armaments race give an adequate return? This undertaking involved, for a country the size of the United States, putting 20,000 firms into action and spending, on space research alone, sums in the order of $5,000 million a year for the purpose of putting a man on the Moon by 1970. Does the impetus given to the American economy and to American science, and the general 'boost' that it gives to the nation, justify all this?

There is still little information in the United States on the problem of defining the 'spin-off' (or 'spill-over' or 'fall-out')—the expressions now used to describe the indirect advantages of the large Federal research programmes. Although disputes have arisen since the launching of the vast space projects in 1958-59, the detailed discussion of the economic effects of the government's action has only recently taken place. Some economists have been alarmed at the size of the expenditure that has been agreed upon for the conquest of space, and have studied the economic return of these researches; industrial interests and Congress have criticized the lack of flexibility and co-ordination in the various systems of patents maintained by Federal Agencies.

Some European economists and politicians emphasize that the action of the Federal Government has influenced competition in favour of United States industry; in other words, the government-sustained economy thus formed in the United States would tend to endanger foreign industries in competition with it, especially those in Europe, by giving United States industries resources and opportunities denied to its competitors.

Information collected in the United States confirms that there are many economic repercussions of military and space research.[1]

1. The results of two opinion polls carried out in American industrial circles are as follows.

In 1963 the *Harvard Business Review* (September-October 1963, p. 14-32 and 173-90) put questions to 13,000 people occupying posts of managerial responsibility in industry. Of the 3,515 people who answered, 71 per cent expected to use techniques derived from space research in their business and 25 per cent said that their business was already using them. [*continued on page* 182

American studies have introduced an important distinction between tangible 'spill-overs' and invisible 'spill-overs'.

The *tangible spill-overs* are the direct effects of space technology in the commercial sphere. They are easily recognized, and their results could be easily measured. They are seen in the arrival of products, equipment and materials which are the results of research financed by the government and used later for private or commercial needs, either in their original form or more usually after alterations and further research work.

The most important of these are the computers.[1] They are followed by a whole list of scientific instruments, miniature electrical components, integrated circuits, control and guidance systems and new materials. These products of government research become component parts of articles sometimes used in everyday life, but more often in industry, medicine or science in general.

The effects are also seen in methods of manufacture, packing and shipment, and even in techniques of organization,[2] which are either applied in their original form or after additional alterations to convert them to civil uses.

The *invisible spill-overs* consists chiefly in the transfer of scientific and technical information. These are used by business firms for their own profit. Often the transfer takes place within the same firm, between its government work and its private projects.

The transfers can be set out as follows.

Transfer of 'know-how'. From this, university research workers and industrial firms derive experience, and a theoretical and practical insight into spheres of knowledge neighbouring their own. In the United States, research workers and technologists pass easily from one sector to another, from one firm to another, and from one university to another. It is certain that the latest 'know-how' spreads rapidly over all the country thanks to this mobility.

Scientific discoveries. Research patronized by the government in the United States has opened up new roads in the sphere of theoretical knowledge, in oriented fundamental research, and in applied research. This is particularly so with, for example, semi-conductors, integrated circuits, energy sources (atomic reactors, solar batteries), high- and low-temperature physics, non-

Another opinion poll was carried out by the review *Industrial Research* (Neil P. Ruzic, 'The Case for Going to the Moon. Part VII: The Case for Technological Transfer', *Industrial Research*, March 1965, p. 67-87). Its conclusions were that the direct and indirect advantages were very important. The same conclusion was reached by an opinion poll carried out by J. G. Welles at the University of Denver, Colorado (Welles and Waterman, 'Space Technology. Pay-off from Spin-off', *Harvard Business Review*, July-August 1964).

See also: *Commercial Profits from Defence-Space Technology*, edited by W. R. Purcell. Boston, The Schur Company, 1964.

1. Military programmes played a more decisive role in their development than programmes of space research.
2. For example PERT (Program Evaluation and Review Technique) sprang from the extreme complexity of the Polaris programme for the United States Navy.

ferrous and rare metals, automation, and the vast field of applied nuclear physics.[1]

Improvement in the quality of products. The demands of the nuclear and space programmes are very searching and aim at ever-continuing improvement of the security factor. The necessity for complete reliability in space, and for the avoidance of nuclear pollution, have produced in some industries a tradition of quality and mathematical precision which brings immediate profit to production for civil uses. Apparatus or equipment which would not have been built—except for a special order—can now be bought straight-away as mass-produced articles.

Reduction of costs. The maintenance by the government of research plant, which can also be useful for industrial research, allows the firms which benefit from contracts to put products on the market whose development and commercial exploitation would have been much more hazardous, or even impossible, if their selling price had had to include the expenses of the research. The large firms, assured of a permanent flow of contracts, have thus in fact had a part of their expenses for civil research paid for by the government, and this allows them to devote more of their money to their commercial transactions.

The value of the invisible advantages is difficult to assess. Some authors[2] believe that it even exceeds that of the direct advantages.

General consequences of subventions and research contracts

A regular stream of contracts and subventions—which tends to become institutionalized—will gradually make industrial enterprises and universities dependent on government administration. It often happens that the whole of the scientific activities of centres supported in this way are gradually shaped to receive contracts or subventions. This is particularly true when the normal financial resources are insufficient, to the extent that the very existence of the centres and research teams depends on their continuing to be financed by contract.

---

1. Owing to the fact that Europe has produced an effort comparable to that of the United States in the uses of nuclear power for peaceful ends, the effect of government contracts is less apparent in this field than in others, where the contrast is clearly visible. One must try to imagine what would be the level of knowledge in this field if there had been no government programmes on either side of the Atlantic.
2. See: John G. Welles *et al., The Commercial Application of Missile/Space Technology,* Denver Research Institute, University of Denver, 1963.

    J. S. Parker, 'Space Technology's Potential for Industry', *Fourth National Conference on the Peaceful Uses of Space.* Boston, May 1964.

    Neil P. Ruzic, 'The Case of Going to the Moon. Part VII: The Case for Technological Transfer', *Industrial Research,* March 1965, p. 67-87.

    Edward B. Roberts, 'The Dynamics of Research and Development'. Address to the National Security Industrial Association, 3 November 1965, p. 11-19.

However, it is in the very nature of this kind of financial aid to be discontinued at some point. The government may hesitate to take the responsibility of terminating such aid for various social or political reasons, although there may be no advantage in continuing and no obligation to do so. But in this way a *status quo* gradually becomes established, in which failure to renew the contracts and subventions becomes more and more difficult. What at first was only occasional support becomes *de facto* institutionalized financing.

Some academic circles[1] even fear that indirect financing—which in the United States comes to 25 per cent of the whole of the resources of the universities—could shackle or compromise their academic freedom.

Finally, financing by contract causes insecurity of employment for the research workers. Industrial firms, and more especially the units and centres of university research, are reluctant to accept firm commitments *vis-à-vis* their research personnel for periods extending beyond the end of the government research contract. So one sees, after renewals of contract, research workers who are getting old without having an established career. True, it is normal that a large proportion of research workers should finish their professional life in other scientific or technical or administrative activities. Older research workers often switch to the professions of teaching or consultancy, managerial jobs, the inspectorate or to executive positions in industry. It is absolutely necessary that the same prospects should be open to research workers recruited to carry out government-sponsored research projects, whether under contracts or through subventions. Very often, however, these careers are barred to them by internal regulations controlling the personnel of the firm, the centre or the university. The reverse should be the case; such careers should be guaranteed for them.

In spite of this, it must be conceded that the scientific, economic and social balance-sheet of subventions and contracts shows a credit balance, although the problems that they raise are by no means negligible.

OTHER SPECIAL STIMULI: SCIENTIFIC PRIZES AND RESEARCH FELLOWSHIPS

The *scientific prize* is a traditional form of encouragement for scientific research. It is a reward for completed scientific work, and usually it crowns the career of a research scientist who has made his name.

The *research fellowship* (or post-doctoral fellowship) is awarded at the beginning of a scientific career, and its object is to finance the training of young research workers who have not yet reached the stage when they can assume the responsibilities which accompany posts such as those of director of research, or professor. It is a necessary adjunct to the *doctoral fellowship,* with which it is sometimes confused.

Other measures can be assimilated to these methods of promotion, such as, for example, subventions to training courses for foreigners, or grants for

1. '*The Support of Medical Research*': *Conference Organized by the Council of the International Organizations of Medical Sciences, London, 4-8 October 1954;* Oxford, Blackwell Scientific Publications, 1965.

the reception of foreign experts, for the organization of scientific conferences, or for the publication of scientific works.

ORGANIZATIONS FOR THE FINANCING OF RESEARCH

In the exercise of their functions of promoting and financing research, certain institutions play a most important part; these are the institutions which have control over the distribution of contracts, subventions and grants for research.

In Chapter 5 we considered the historical circumstances in which these institutions were founded in certain countries.

The study of the scientific structures in the different countries brings to light considerable varieties in these institutions[1] as to their status, their objectives, their activities and their management. They are called by different names: 'central organizations of research', 'organizations for the promotion of research', 'national centres of research', 'research councils', etc.

Their status can be public[2] or private.[3] In the latter case they are often organizations whose character of public utility has been recognized by the government, such as the philanthropical foundations.

Their powers may be general, that is they may extend to every branch of science (for example: the National Science Foundation in the United States, CNRS in France); or they may be restricted to one branch of science (the Research Councils in the United Kingdom, Norway and Sweden); or they may be directed towards some kind of organizational pattern of fundamental or technological research (ZWO in Holland, FNRS and IRSIA in Belgium, the Council of Scientific and Industrial Research in Norway, Council of Technological Research in Sweden, etc.).

In the countries where the organizations are divided among various sectors, this appears to be the result of the play of forces sometimes exerted by the scientific community itself, and sometimes by the government in pursuit of its own science policy; they tend to favour certain scientific sectors which thus becomes assured of preferential treatment. In a large number of countries the result of dividing the organizations among sectors has been to confer a privileged status on the sciences of biology and medicine, the agricultural sciences and the nuclear sciences. Besides, the more these form a closed

---

1. On this subject consult: *World Directory of National Science Policy-making Bodies* published by Unesco: vol. I: *Europe and North America* (Paris, 1966); vol. II: *Asia and Oceania* (Paris, 1968); vol. III: *Latin America* (Paris, 1968); vol. IV: *Africa and the Arab States*, is to appear in 1971.
2. For example: Institut pour l'Encouragement de la Recherche Scientifique dans l'Industrie et l'Agriculture (IRSIA), in Belgium; Centre National de la Recherche Scientifique (CNRS), Centre National d'Études Spatiales (CNES), Direction de la Recherche et des Moyens d'Essai (DRME) of the Defence Minister, and the Commissariat à l'Énergie Atomique (CEA) in France; the TNO and the ZWO organizations in the Netherlands; the Research Councils and Atomic Energy Authority in the United Kingdom; various Federal Agencies in the United States, e.g. NASA, NSF, NIH, AEC, etc.
3. The Deutsche Forschungsgemeinschaft in the Federal Republic of Germany; the Fonds National de la Recherche Scientifique (FNRS) in Switzerland, etc.

circle, the more their leading figures tend to become overpowering. This system of scientific organization, which often discriminates against junior research workers and new scientific fields, is less noticeable when the organizations for the financing and promotion of research have general powers than when they are confined to one scientific or economic sector only.

The general function of financing and promoting research is sometimes shared with the responsibilities of performing *intra muros* research and even tasks of a more general kind, such as performing scientific services. Such combinations of responsibilities cause harm because they introduce the risk that the support given by these financing organizations to other *extra muros* research projects may be too much influenced by the needs and opinions of those in charge of their own internal programme. The temptation may crop up in this case to consider the first as merely ancillary to the second. A definite separation of the functions of financing research from those of actually performing research has the advantage of making a clear distinction between the different spheres of responsibility, and in the end guarantees greater objectivity in the choice of research projects and in the selection of research personnel.

Two common features can be distinguished from the diversity of national organizations: the financial dependence on the government of the organizations for the promotion of research, and the freedom of action they enjoy (which is often very wide). It must be pointed out that even the organizations which have kept a private status derive at present the larger part, if not the whole, of their resources from the government. This is the result of a historical process: as the needs of science increased, the private contributions to these organizations, and the income from their endowments, became insufficient. The government has taken over the task of private finance.

This evolution means that the distinction between these institutions according to their status has become purely formal. Whether they are public or private, their essential role is usually to distribute among the research workers, teams, and centres, an agreed part of the sums of money which the government apportions to the financing of research.

These institutions usually enjoy wide powers of decision, regardless of questions such as their status (public or private), whether or not their directors are nominated by the government, and whether these directors are administrators or scientists.

The freedom which these organizations possess in deciding the destination of public money is not always offset by proper government supervision. Here we are not dealing simply with financial supervision, which is usually carried out most correctly, but supervision to ensure the best use of the resources entrusted to these institutions by the nation. Sometimes it has been necessary for younger research workers and those who are promoting new sciences to draw attention to the old-fashioned methods of administration which have often congealed the power structures. Sometimes governments have grown anxious at the tendency to renew the same subventions continually, and at the unsatisfactory career opportunities and working conditions of the research workers in the subsidized programmes, and have accordingly sub-mitted the management of these organizations to a close scrutiny. This is a delicate problem, because of the established positions and the widespread fear—both in the scientific community and within the governments—of the prospect of more direct government intervention degenerating possibly into amateurish meddling in scientific work itself.

The power structures which have been formed through the perpetuation of the authority of administrators and experts attached to these organizations recall in many ways the 'boss system' of some university faculties. The young intellectuals often have to mark time before obtaining admission into the power group; and when they have arrived, they 'boss' their successors. The very rapid extension of the base of the pyramid of research workers clearly causes critical tensions in this somewhat archaic structure.

However, in several countries, things appear to be on the move; there is a new interpretation of the relationship between these organizations and the government, based on clearer policy directives as to the allocation of resources and more active supervision of the way they are handled. The composition of the decision-making committees in these organizations is also changed more frequently. But the principle of their autonomy of action is zealously respected, because direct intervention by Ministers and the govern-ment bureaucracy in 'microdecisions' about personnel, and about research projects, has rarely produced good results.

There is now a tendency to make the power structures of the organizations less rigid by means of more rapid changes in the membership of their administrative and scientific boards, and by the introduction of scientific criteria in the preparation of decisions, thus reducing the range of arbitrary choices and increasing responsiveness to considerations of general national interest. In such a frame of operation, the government can make known the total quotas for the various fields of science and the various kinds of re-search, while at the same time specifying certain priority options. The administrative and scientific boards of the research foundations, the research centres or the national research councils would thus, while remaining free to make their own decisions, contribute more towards the realization of the national science policy.

It is also felt that to avoid the dangers of 'crystallization', and in order to make funds constantly available to finance fresh initiatives, these organiza-tions ought to restrict themselves to promoting the launching of research programmes which, because of their novelty or the cost of the apparatus required, do not yet have a place in the normal activities of a university, research centre, or industrial enterprise. The responsibility of financing and managing a research project ought therefore to be transferred as quickly as possible to the institution where the project is being carried out.

These organizations would then play the role in scientific development which the banks and finance corporations play in the economy, allowing the most dynamic and the most imaginative groups to anticipate on the growth of their regular budgets in order to put new projects into operation.

The role of the planner is to direct the bulk of the money towards *tasks* which are most advantageous to the national development. The role of financing and promoting organizations is to apply the policy of the plan to particular cases, by only granting funds to those who deserve trust. In this way the managers of these organizations will be left that degree of freedom without which there is no responsibility. But they must exercise it in conditions of independence and objectivity, and these conditions must be such that the prospect of emigration does not attract away young and talented intellectuals. There must be no barriers raised by a power group or a 'boss', or any other out-of-date array of influence permanently controlling the study of projects and the decisions to grant funds.

FOURTH FUNCTION:
CARRYING OUT RESEARCH

The objectives have been determined, the research projects have taken shape, the money has been made available. The problem now is to bring the projects to fruition—to equip the units of research, to provide them with research workers, and to organize their activities.

It does not enter into the scope of this work to deal with the management and direction of research itself. This is a field of study in its own right, and there are already numerous books and articles on the subject, written by heads of research units who have acquired considerable experience, by managers of research in large industrial enterprises, and by directors of research institutions.

Our comments will therefore go no further than the threshold of this new subject—that is to say to the exact point where the responsibility of the government authority ceases and the responsibility of the director of research begins. This point is the choice of the institution, laboratory or research unit which is to have the task of performing the research.

THREE SECTORS

A distinction is usually made between three sectors where research is performed: the government sector, the university sector, and the sector of industrial enterprises.

The first sector contains the scientific institutions of the government. The majority of them were founded to fulfil some special scientific need (such as meteorology, geodesy, hydrology, etc.); but in recent times the governments have founded several centres whose principal activity is research and experimental development. The function of public services is, however, always implicit. In this first category will also come non-profit-making private institutions when the bulk of their finance comes from government sources.

The major objective of the second sector is education and training. But from the nineteenth century onwards, academic research has found its chosen ground in the university, and this truth-searching activity of the intellect has clearly become indispensable for ensuring the quality of higher education. For this reason it is considered normal and necessary in most countries for the staff of universities to divide their time between teaching and research.

The primary objective of industrial enterprises is profit-making production. Their part in carrying out research (which becomes very large during the second phase of industrialization) derives from the need to invent products and processes on which their future production will rely. The industrial research associations are usually included in this third sector, although they occupy a position half-way between this sector and the first, since their activity should benefit the nation as a whole and their financing is ensured—at least in part—by the government.

A special place must be given to research enterprises (which endeavour to derive their incomes from the fees which industry pays for the results of their research) and to the contract research centres. These organizations would form a fourth sector if research tended to become a profit-making activity on its own account as a viable commercial proposition. It seems likely that, however beneficial these centres are, they will continue to rely permanently on a large volume of government contracts, and that no government has made them an essential part of its national organization.

PARTICULAR NEEDS OF EACH SECTOR FOR CARRYING OUT RESEARCH
AND EXPERIMENTAL DEVELOPMENT

In general, the institutions belonging to the three above-mentioned sectors do not have as their primary and explicit goal the advancement of knowledge or the invention of new methods and plant. The exception is the research carried out in the government sector (especially in those units where the functions of public service are not very important) and in the contract research centres.

However, the units of each of the three sectors have special motivations and needs as regards the carrying out of research and experimental development, for there is a minimum amount of activity of this nature without which the institution or enterprise would lose its professional skills in its primary function. This minimum is essential, even if transfer of scientific information or transfer of technology from foreign sources are relied upon to satisfy the main needs.

The industrial enterprise which does not carry out research and which therefore has to pay for all its technology—in royalties for example—is in danger of gradually losing the basic knowledge of its business and of becoming incapable of making the right choices when buying licences. When this happens, it has no alternative but to be taken over. Similarly, agricultural extension units of the government which do not to carry out enough research will undoubtedly cease to be properly qualified in their technical assistance function to the farmers. Finally, a university department which does not base its teaching on research will gradually slide down to the level of a professional school.

In short, there is a minimum level of research activity to be maintained in the organizations belonging to each of the three sectors. But these organiza-

189

tions can usually increase the volume of their research activity far above this minimum if they succeed in mobilizing the necessary resources.

A prime duty of the planner will be to ensure that the minimum level of research activity is within reach of the institution's own financial resources— that is, its regular budget if it is a government institution or if it belongs to the education sector, its gross profit if it is in the business sector, and its subscriptions if it is an industrial research association. If the ordinary resources of the institution do not suffice and cannot be increased, extra funds from the government may be necessary.

CHOICE OF SECTOR BY THE PLANNER

When the necessary minimum level of research has been obtained in the three sectors of the nation's research and experimental development plant, the choice still has to be made as to where those parts of the nation's research programme which have not yet been allocated will be carried out. Some countries, like the United States, have entrusted the main share of such research and experimental development work to private enterprise. Others have entrusted the greater part to government centres, which they have possibly founded for this purpose. Others again have put the universities in charge, right to the limit of their possibilities.

There is probably no simple answer to the question: 'Which is the right way?'

The approach which we wish to suggest here, which is based on studies made in several countries, postulates a special aptitude for each of the three sectors where research is performed.

The special vocation of industrial enterprise is *production*. If, therefore, the primary aim is to invent new appliances and equipment in order to produce them as an economic proposition for the needs of the population, and afterwards to get the highest commercial value out of all the 'spill-overs' both on the internal market and on the world market, then the best choice would be industrial enterprise, even if some fundamental research had to be carried out. It is, in fact, likely that a large number of results obtained in government centres or universities are still inadequately exploited, because it is not the function of anyone in these institutions to think about commercial benefits or industrial applications. On the other hand, an industrial enterprise will be concerned about this, because production is its main function, whatever the customer or the use. This is perhaps why the spread of computers in business management has been much more rapid in the United States than in the U.S.S.R., although in both cases this is a 'spill-over' from military and space research.

The special vocation of the government sector is *continuity of service*. It would be improper to entrust meteorological research to an industrial enterprise, because the value of this research depends on an increasing accumulation of data and exact measurements, some of which appear to have no immediate object. An industrial enterprise has no aptitude for indulging in this kind of work. It would be equally improper to rely on a university for such work, for no educational institution would wish to repeat the same routine observations for decades to come.

The special vocation of the university is to direct the *intellectual curiosity* of its research workers towards unusual aspects of nature or scientific speculation—those aspects which pose unresolved problems. It must keep itself in the forefront of science without having to concern itself with anything but education in the widest sense of that term. Unlike private enterprise, the university does not aim at pursuing a technological breakthrough right to its profit-earning stage (especially if the difficulties which remain to be overcome are merely technical ones). And by the same token, unlike government services, it does not aim at carrying the burden of routine science—with the preoccupation with methodology and with the continuity of observations or measurements which that implies.

Whatever the primary vocation of an institution or enterprise may be, one has to face the following problem: too much research can disorganize production, or education, or the government service, by diverting the organization away from its basic objectives.

There have been examples of firms collapsing at the end of government research and experimental development contracts; they had ceased to be efficient producers and sellers. There have been examples of government scientific services becoming disorganized by being given too much money to spend on research. University professors have neglected their teaching to satisfy the demands of defence authorities or space research administration, with which they had signed contracts.

We do not wish to attach too much importance to this argument. An efficient administrator can more easily adapt his organization to a surplus of resources than to a lack of them. If a subordinate activity encroaches too far, it is relatively easy to detach it by creating a special department or a separate administrative structure with which the main unit will keep in close touch without being burdened by it.

At all events, it appears that the special vocation of each of the organizations belonging to the three sectors of research performance can be used as a guiding line to choose the sector where the research is to be carried out.

8 International co-operation in science

As yet there is no international policy for science, in the sense of a common line of policy in scientific matters such as might be laid down by a certain number of States. There is, in fact, no authority competent to lay down such a common policy and to make the decisions which become urgent for all the partners in an association.

But international scientific relations have evolved in such a way that the national governments are forced from now on to take, in concert, decisions that relate to co-operation in certain fields of science and technology.

International scientific relations have developed considerably since the nineteenth century, and they have taken many forms. This development has been accompanied by a growing measure of intervention on the part of governments.

HISTORICAL SURVEY

'Even less than other kinds of knowledge', writes A. King,[1] 'can science be confined within national boundaries. The structure of chemical compounds is happily the same in every country, no nationalistic doctrine has yet been able to change the laws of gravity, radioactive fall-out remains uninfluenced by political considerations, and the scientific approach can be applied to the study of all kinds of systems. We have been the unhappy witnesses of the excesses of nationalism, and we know that every effort to fit scientific truths to political opinions ends up by foundering in ridicule.'

Science is therefore more than a common heritage of humanity. It is also, as we have seen, a privileged sphere of relations where intersubjective agreement among men is achieved.

Since ancient times, and more particularly since the Renaissance, the spread of scientific ideas and methods has always surmounted frontiers, often even when the nations concerned were at war. Scholars, both because of

---

1. A. King, 'International Scientific Co-operation—its Possibilities and its Limits', *Impact*, vol. IV, no. 4, 1953, p. 189.

their ideals and through necessity, have always had the 'international out-look'.

Science inside any country has always been sustained by foreign intellectual contributions, and if at times nationalism has raised barriers between scientists, it was for short periods and accidental reasons.

CO-OPERATION BETWEEN SCIENTISTS

The most ancient form of international scientific relations is a completely spontaneous and unorganized form, consisting of personal intercourse or correspondence between scientists who wish to exchange their publications and compare their methods and the results of their research. In the times when research still found no place in learned societies or universities, these individual exchanges helped to create an extensive intellectual community, which often supplied moral if not material support to the man of science who was exposed to political or philosophical constraints in the society in which he lived.

This kind of intercourse has preserved its vitality today, and is undoubtedly at its most effective on the occasions of the innumerable conferences, meetings and congresses which promote cross-fertilization of ideas.

When the scientific societies were formed, personal relationships were reinforced by more permanent relationships. Throughout the eighteenth and nineteenth centuries, the academies of science and the scientific societies established contacts with each other, which, sometimes at least, gave rise to certain concerted programmes in the fields that particularly lent themselves to international co-operation, such as astronomy, meteorology and geodesy.

It was during the second half of the nineteenth century that the process of institutionalizing international scientific co-operation began. In 1914, there were already about fifty international scientific organizations.

According to M. Daumas,[1] the first international scientific organization was the Geodesic Association of Central Europe, founded in 1861 in Berlin to co-ordinate the geodesic operations undertaken in Germany and in the bordering countries. After some years of activity the association invited other countries to join; in 1896 an international convention was signed forming an International Association of Geodesy in which twenty-three countries were represented.

At the end of the nineteenth century there appeared the first specialized international scientific unions. As early as 1900 the International Association of Academies attempted to co-ordinate the activities of these various organizations, but it was thwarted by the political troubles of the years which preceded the First World War.

After that war the association disappeared and was replaced by the International Research Council. 'The new organization had as its aim to co-ordinate international activity in the different branches of science, to

---

1. Maurice Daumas, 'Esquisse d'une Histoire de la Vie Scientifique', *Histoire de la Science*, p. 170. Paris, Gallimard, 1957 (Encyclopédie de la Pléiade).

encourage the formation of international associations of unions, to direct scientific activity in the fields where there were no international organizations and to enter into agreement with the governments of the member countries to encourage the study of questions that were completely within its competence.'[1]

The International Research Council was replaced in 1931 by the International Council of Scientific Unions (ICSU), which played a very important part in the formation and development of international scientific unions between the two world wars. After the Second World War, ICSU was recognized by Unesco as the international organization qualified to represent scientific unions and to co-ordinate their labours.

CO-OPERATION BETWEEN GOVERNMENTS FOR SCIENTIFIC ENDS

It was also at the end of the nineteenth century that governments began to take an interest in scientific co-operation. Political considerations, though still insignificant, played some part, as well as purely scientific considerations. The formation of the International Bureau of Weights and Measures in 1873, following on a recommendation of an international commission of scientists, was the first sign of this new attitude of political authorities.

But at this period, as J. J. Salomon writes,[2] the role of governments in this field 'was still very circumspect', since it was confined to clearing the way, at government level, for the establishment of international links between scientists of different countries. In short, they were content to lend their services, and if they intervened, it was during the last stage of the negotiations themselves. It is easy to see the reason for this: in the nineteenth century science did not yet involve any of the major interests of the governments, and the organizations which were the subject of government agreement had such insignificant objectives and budgets that they did not pose any important political problem.

This view was modified a little after the First World War. Governments had been made aware of new vistas opened up by science for warfare. Science gradually became considered as one of many instruments of international policy. Attempts were made within the framework of the League of Nations to start scientific organizations. But these attempts did not lead to any results in the sense that they did not pass the stage of purely cultural co-operation.

It was only after the Second World War that governmental attitudes assumed clearer shape. Their interventions followed each other so rapidly, that there are no longer today any important scientific agreements where the governments do not at least hold a watching brief.

Three factors intervened, according to P. Auger:[3] the replacement of the League of Nations by the United Nations, which had much larger resources

1. A. King, op. cit., p. 200.
2. *Minerva*, London, 1964.
3. Pierre Auger, 'Scientific co-operation in Western Europe', *Minerva*, vol. I, no. 4, summer 1964, p. 431.

at its command; the interest shown by the two great powers, the U.S.S.R. and United States of America, in the International Council of Scientific Unions; and lastly the efforts made towards European unification.

In this way, directly after the war, large scientific and technical institutions, specially constituted for an international role, were formed or grouped together under the aegis of the United Nations (Unesco, WHO, WMO, IAEA, UNIDO, etc.) which perform activities closely linked with their general assignment.

Other intergovernmental organizations with regional powers, such as NATO, Comecon, the OECD, the OAS, the OAU, developed scientific activities to different extents within the framework of their general assignment, of a military, political or economic kind.

In the European sector one has seen the creation of organizations whose essential or even only task is research: CERN in 1953 (under the aegis of Unesco); Euratom in 1958; ESRO and ELDO in 1962; EMBO, founded in 1963; and the European Conference on Molecular Biology, ECMB, the agreement for which came into force early in 1970.

It may be said therefore that the evolution of international co-operation has followed three main trends.

The first phase showed a trend towards the institutionalizing of the relations between men of science. From being personal and casual, as they still were in the nineteenth century, they became progressively more organized and permanent within the framework of specialized organizations which brought together the scientists of several countries or the whole world.

In the second phase the co-operation, which originally had purely scientific objectives, became more and more influenced by political and economic motives. Governments have realized the advantages which science offers them both in their internal affairs and in their foreign policy. In this way a certain number of intergovernmental organizations have sprung from the governments' concern, which is primarily political, to enter into a partnership by means of scientific co-operation. Such partnerships are composed of a number of States linked by a common understanding in economic and political matters. The case of Euratom can be quoted as an example; its formation falls within the framework of the efforts for European unity, and its aim is to ensure the independence of the production of nuclear energy for six European countries. The formation of European space organizations, such as ELDO and ESRO, was in response to political considerations of the same kind.

Most often the political considerations are in line with the opinions expressed by the scientific community, who thus enjoy support from the diplomatic action of the governments. However, in some cases political motives can be the determining factor and go far beyond the moderate enthusiasm, or even reluctance, of the scientific world.[1]

In other cases political motives have been opposed to the wishes of the scientific community and have thwarted projects proposed by the scientists.

1. This was the case of the International Cancer Agency, created in 1963 at the instigation of the French Government; five European countries subscribed to it.

This was the fate of the project for the creation of an institute of advanced technology within the framework of NATO, and of the project for an international biomedical laboratory within the framework of the World Health Organization, which was rejected in 1965.[1]

CO-OPERATION BETWEEN GOVERNMENTS IN THE PRESENT PHASE

In the third phase, in which we are at the moment, international co-operation is becoming at the same time both scientific and technological. There are two reasons for this. The first is that the international scientific programmes call upon the co-operation of industry by giving them orders for the construction of their scientific infrastructure, which consists of equipment, apparatus and a quantity of instruments which are required for experimental work. This is particularly the case of two European organizations for fundamental research. CERN made use of industry at its inception to construct its particle accelerator and more recently to build a storage ring; likewise, the project for a large accelerator of 300 GeV, if it should succeed, would clearly demand considerable participation by European industry over a long period. ESRO had ordered several pieces of apparatus for the construction of its laboratories and its scientific satellites.

The second reason is that a number of scientific programmes are undertaken with the idea of a practical application. Such is the case of ELDO, whose assignment is to construct rockets to launch the European satellites, and of Euratom and ENEA, whose assignment is the preparation of prototype nuclear reactors. The existence of common objectives, reaching beyond the purely scientific phase, is conducive to the formation of an integrated common policy from the research stage right to the stage of industrial exploitation and commercialization. We shall speak of this again later when we mention the problems which international scientific co-operation poses.

FORMS OF CO-OPERATION IN EXISTENCE TODAY

The forms of international scientific co-operation in existence today are highly diversified. We shall try here to draw up a typology of the forms of international scientific co-operation in existence, both bilateral and multilateral.

To provide an outline. a classification can be adopted according (a) to the function fulfilled by co-operation; (b) to the legal status of the organizations; (c) to the geographical area in which co-operation takes place; (d) to the scientific field covered.

1. See Hilary Rose, 'The Rejection of the WHO Research Center', *Minerva,* vol. IV, no. 3, spring 1967, p. 340.

CLASSIFICATION ACCORDING TO FUNCTION FULFILLED

Exchange of information

As far as the simple exchange of scientific and technological information is concerned, interchanges are free and spontaneous. Responsibility for them rests on the scientific world alone, and governments take no part though they may give certain financial support, and assistance in the duplication of documents and information, and even give official support to the co-operation by means of a bilateral cultural agreement.

In general, exchange of information does not require that, before starting research, the research workers should first meet or agree on concerted action. Co-operation is achieved *ex post facto,* and depends on the dissemination of information about research in progress and research that has been completed in the participating countries. This kind of co-operation takes the form of exchanging publications and of conferences, where the research workers discuss together their ideas, their methods of work and their results.

At the dawn of the automatic processing of information and especially of bibliography, co-operation by the exchange of information faces the prospect of making a new leap forward, which will entail important international agreements; Unesco is already working on these.

Co-ordination of national action

Co-ordination may take the form of concerted research programmes which include provisions for exchanging the results of research. The research workers meet to share out the work according to a flexible programme and agree to exchange and compare the results of their work. This co-operation sometimes entails a standardization of methods.

One of the aims of this co-operation may be to ensure that a scientific field is completely covered, that is, to make an inventory which is as exhaustive as possible (International Geophysical Year; International Hydrological Decade), in the fields where each research worker depends upon the existence of a regional or even world-wide scientific survey for the conduct of his own research. For example, Unesco has particularly concentrated its energies on research which is concerned with the sciences of the human environment, on a world scale.

Another aim is to avoid unnecessary duplications of the programmes of the different laboratories of research teams (as is done, for example, in the case of teams operating within the framework of the Committee for Research Co-operation of the OECD).

Under this arrangement each participating laboratory or country pays its own expenses, but a wide dissemination of the results is ensured by means of periodical reviews, meetings, symposia, card indexes, etc.

Co-ordination of the national efforts may go even further. It then develops into a common programme in which the work is divided up in advance. Here we are dealing with a single project—a technical whole, of which each country undertakes definite parts. Each participant country pays its share of the expenses, but the work is integrated from the start. An international

197

secretariat or one of the participant countries ensures the administrative and technical co-ordination.

Under this heading can be classified the major international oceanographic expeditions organized under the auspices of Unesco, the programmes of experimental reactors put into operation under the aegis of the European Nuclear Energy Agency, ENEA (Halden, Dragon); the satellite Symphonie (France, Federal Republic of Germany); the prototype of the breeder reactor where the work is shared between Belgium, Federal Republic of Germany and the Netherlands; the Concorde project (France, United Kingdom); the project of the Airbus (France, Federal Republic of Germany, Netherlands, United Kingdom).

Common action expressed in the form of institutions

The common project with a common budget. This kind of co-operation is characterized by the existence of a common programme and a common treasury, fed by national contributions, whose distribution has been decided upon at the start. The common budget is administered by an international organization. The work is allotted among the various countries and there is no central laboratory. The project gives rise to a series of contracts placed by the organization in the different countries which are taking part in the project; these are paid for out of the common treasury. The geographical distribution of these contracts is decided by a deliberative body according to rules laid down by a majority vote, and the constraint of quotas is sometimes provided for in the agreement. ELDO is an example of this kind of co-operation.

The common project with a common research centre as well as budget. The research is carried out in a centre that brings together research workers from all the participant countries. This centre may belong to one of the participant countries and be put at the disposal of the international community or of an international organization (for example, the Von Karman Institute for Fluid Dynamics, in Brussels). It may, on the other hand, be a proper international centre: this is the case of the laboratories of ESRO, Euratom and CERN. Certain international organizations combine the formula of contracts for research with that of a common centre (ESRO, Euratom). They then make a distinction between intra-mural and extra-mural activities.

CLASSIFICATION ACCORDING TO THE LEGAL STATUS OF THE ORGANIZATION

There is no legal definition of the international scientific organization. The *Directory of International Scientific Institutions* published by Unesco (2nd edition, 1953) already mentioned 264 of them; a catalogue published in 1965 by OECD under the title *International Scientific Organizations* listed 381.[1] This catalogue mentions, however, that it is very difficult to reach a moder-

---

1. Consult also, *Yearbook of International Organizations,* 12th ed., Brussels, Union of International Associations, 1969.

ately accurate total, let alone a complete one, of the international organizations whose statutes refer to science in one way or another. In some institutions the scientific interests are dominant, in others the scientific activities are derived from one or several assignments of a more general nature, with an economic, social or agricultural slant, or concerned with health or culture, etc. It would therefore be better to speak of international organizations competent to deal with scientific matters.

It will be useful to make a distinction between non-governmental organizations and intergovernmental organizations.

The non-governmental organizations are formed as the result of private initiatives. They proceed directly from the international scientific community, and are very numerous. The majority of them confine themselves to fostering contacts, and exchanges of information and methods, among the research workers. Only very few direct or concentrate their efforts on a common programme of research, with the exception of ICSU which, in close liaison with Unesco, ensures the co-ordination of programmes carried out by the federated scientific unions within its own organization and the national scientific academies and societies, which are the national members of this organization.

The non-governmental organizations are usually established in the form of international associations and are under the jurisdiction of the country where their legal incorporation has been enacted. They are characterized by the variety and also the flexibility of their regulations, which give them a wide freedom in the choice of their activities and the recruitment of their members. They usually function on limited budgets and small staffs. A large number of them have asked for government financial support, which has either been given to them directly or through the good offices of intergovernmental organizations.[1]

The intergovernmental organizations are those which are established by an agreement between governments. The original idea sometimes comes from scientific circles, but the formation, financing and administration of these organizations are the responsibility of the governments. The list of the intergovernmental organizations with scientific capacities is shorter[2] than that of the non-governmental organizations. But there is certainly a great variety in their composition, organization, procedures and functions.

CLASSIFICATION ACCORDING TO THE GEOGRAPHICAL AREA
IN WHICH CO-OPERATION TAKES PLACE

A distinction can be made between: (a) world organizations or programmes; (b) regional organizations or programmes; (c) bilateral or multilateral agreements or programmes.

---

1. This is the situation of Unesco, which gives its support to a certain number of non-governmental organizations, such as ICSU.
2. The OECD catalogue, already quoted, numbered 56 of them.

World organizations or programmes

Co-operation at world level is the task of the United Nations and its specialized institutions,[1] such as Unesco, FAO, WHO, WMO, IAEA, and UNIDO.
It is also the task of certain organizations which do not belong to the family
of the United Nations, like the International Bureau of Weights and Measures, the International Hydrographic Bureau, the International Institute of
Refrigeration.

Regional organizations or programmes

This form of co-operation brings together on the scientific plane countries
which are already linked by common political or economic objectives and
which share the same social, cultural and university traditions.

It is particularly in Europe that regional co-operation has developed,[2] but
regional scientific organizations also exist in Asia[3] and in America.[4]

Bilateral or multilateral agreements

This kind of co-operation covers a whole gamut of agreements for co-operation in cultural, scientific and technical matters, and for production. They
link together the governments, laboratories, universities, or industrial enterprises of two or more countries.

These agreements can be regrouped in two categories: (a) cultural and
scientific agreements; (b) industrial agreements.

Cultural and scientific agreements. Numerous agreements concluded between
two countries provide for the exchange of research workers, professors,
probationers and lecturers; the organization of seminars and symposia; the
exchange of methods and tests, the exchange of prototypes of experimental
apparatus, etc. This scientific co-operation is often arranged within the
framework of more general cultural agreements, which include 'scientific
clauses'.

Industrial agreements. Co-operation between firms belonging to different
countries may assume various aspects: production under licence; agreements
for technical co-operation; the association of firms in an international
consortium or group; the setting up in common of subsidiary companies or
departments of research; incorporation or taking over.

The co-operation may be aimed at the solution of a particular problem,
or it may be more permanent. In the latter case its effects on the industrial
structure of the countries and regions involved are clearly much more
important.

1. See *Yearbook of International Organizations,* already quoted.
2. CERN, Euratom, ECSC, ENEA, ELDO, ESRO, ECMB, CETS, CEMT, etc.
3. The Colombo Plan, the Asia Pacific Forestry Commission, etc.
4. OAS, Pan American Health Organization, the Institute of Nutrition of Central
   America and Panama, the Inter-American Institute of Agricultural Sciences, the
   Latin-American Center for Physics, etc.

Scientific and technical dealings between industrial firms can be traced when they are accompanied by dealings connected with patents. But exchanges that have a bearing on knowledge relating to the processes of manufacture are rarely made public.

The governments may be associated with these agreements and contribute financially to the project. Sometimes they even take the initiative. Political considerations are then added to the industrial and commercial interests that are at stake. This is the case of the Concorde project for building a supersonic aircraft (France, United Kingdom), and the project for the preparation of the satellite Symphonie (France, Federal Republic of Germany). It is also the case of the Airbus project which could lead to a joint venture involving France, the Federal Republic of Germany, the Netherlands and the United Kingdom.

On the other hand, the new links uniting the governments and the large industrial enterprises of Europe, at least in certain industrial sectors, as well as the decisive importance of government orders to establish the first buyers' market of a new product, are likely to give rise to types of new international schemes and agreements.

CURRENT BALANCE SHEET FOR INTERNATIONAL SCIENTIFIC CO-OPERATION

EXTENT OF CO-OPERATION

Activities in the sphere of international scientific co-operation have been considerably increased during the last ten years, and so have budgets of international organizations.

As an example one may cite the budgets of certain European organizations in 1968; the figures are, in millions of dollars: CERN, 60; Euratom, 88; ELDO, 97; ESRO, 50; that is, a total of 295 million dollars.

On the national level, international co-operation disposes at the moment of very large financial resources in many countries. This is especially the case in Western Europe, where the expenditure on international activities represents, according to the countries concerned, from 10 to 50 per cent of the total resources which governments devote to research.

The weight of financial contributions towards international co-operative programmes can therefore become very heavy for national budgets.

MOTIVES FOR CO-OPERATION

The motivations in co-operation experiments which have taken place so far are clearly varied. It is, however, possible to classify them in three principal categories.

First, co-operation may result from the necessity of employing such considerable material resources and numbers of research workers that these are beyond the capabilities of any one country. To reach the threshold of efficiency it is absolutely necessary to regroup resources which can be supplied by the different participating countries (example: CERN).

Secondly, certain fields of research (for example oceanography and meteorology) are naturally incapable of being limited by national horizons and can only be explored systematically at an international level.

The third motive, which is also the newest, is part of the prospect of science-based development. It springs from the need to undertake certain scientific activities in order to achieve objectives of greater scope whose prosecution needs joint efforts.

Thus certain international organizations with an economic *raison d'être,* such as the European Economic Community, are starting programmes of technological research in some specialized branches of industry, and organizations that specialize in economic and social development have sponsored research on health or food (WHO or FAO). Here, scientific objectives have become part of operational objectives.

However, the reasoning underlying scientific and technical co-operation has hitherto been less premeditated, or at least less precise, than the foregoing analysis might suggest. Very often, diplomatic or political considerations have been predominant.

ORIENTATIONS OF CO-OPERATION TODAY

Six principal orientations of co-operation can be distinguished, as set out seriatim below.

Fundamental research and oriented fundamental research, in which co-operation in the form of international scientific conferences and international scientific unions is already of long standing. Among the institutions of the United Nations, Unesco is the only one to have a general assignment in this sphere.

The environmental sciences: oceanography, hydrology, geophysics, the sciences of the atmosphere. Intergovernmental organizations are numerous in these fields. Some of them date from the nineteenth century (International Meteorological Committee, 1872). Some of them are regional, for example, the International Commission for the Scientific Exploration of the Mediterranean (eight European member States). At the world level, Unesco and the WMO at the moment play a decisive role in scientific research concerned with the human environment.

Technology and the specialized branches of the applied sciences, where a large number of organizations seek both to promote one particular sector of the economy and to improve its technology. Within the United Nations system, the Food and Agriculture Organization (FAO), and the United Nations Industrial Development Organization (UNIDO), are examples of this. One can also quote the activities of the Committee for Research Co-operation of OECD, as well as the work of the permanent industrial branch commissions of CMEA (Comecon). There is a very large number of scientific or professional non-governmental organizations, whose field of specialization is narrow, in particular when it is confined to a single agricultural or industrial product or to a specific problem such as the protection of the cultural heritage, or pollution.

The scientific public service activities, without having a primarily innovative character, nevertheless prove to be a fertile ground for international co-operation, for instance in the sphere of weights and measures, for the study of which an International Bureau (IBWM) was set up in 1875. There is also the International Organization of Legal Metrology, the International Hydrographical Bureau, and also the unions and associations concerned with documentation. Very important activities, like analytical chemistry, the inspection and standardization of industrial products or food, can be classified in this category or in the preceding category; the organizations concerned with them play an important part in effecting liaison between different disciplines and between scientists and technicians in branches in which common auxiliary techniques are used.

Advanced research and technological development are to be found mainly in space research (ELDO and ESRO) and atomic physics (Euratom, CERN and ENEA).

The medical and biological sciences: in this field co-operation is growing rapidly in the fundamental biological sciences. The same thing can be said of the sciences of food or certain environmental sciences.

The Organisation for Economic Co-operation and Development estimates that out of the total financial resources allotted to international organizations for scientific co-operation by the member countries of that organization, 50 per cent goes to nuclear research and 30 per cent to space research. This makes it possible to say that the greater part of the international co-operative effort in Western Europe is today concentrated on 'big science' and 'big technology'.[1]

PROBLEMS OF CO-OPERATION TODAY

The multiplication of international scientific programmes, and the heavy expenses which therefore fall on national budgets, have raised in the last few years a number of problems at government level.

The recent history of European scientific co-operation in space research and atomic physics makes it possible to assess the difficulties. *Mutatis mutandis,* the example of Western Europe can no doubt be transposed to a larger scale, and the lessons drawn from it seem to be valid everywhere.

Example of space research in Western Europe

European interest in this field dates back to April 1960 when a group of space scientists met in London and made proposals for a European space programme.

1. The journalistic expressions 'big science' and 'big technology' allude to the unprecedented size of the apparatus and budgets which have become necessary in a small number of sectors of fundamental science and in the ultra-modern sectors of technology.

The work of the group resulted, at the end of 1960, in an intergovernmental conference, held at Geneva, with the object of signing a convention between a number of European countries and setting up a Preparatory European Commission for Space Research. The Commission's assignment was to work out a programme and to formulate draft statutes of a European organization for space research.

This organization (ESRO) was created by a convention signed in Paris in June 1962; its objective was to make available for European research workers a technical infrastructure which would allow them to carry out scientific experiments either with small-sized exploratory rockets, or with satellites. Ten European countries took part. The chief aim of co-operation was to create an international and unified logistic support for such experiments. But since in this field equipment and apparatus must often be built, technological development became an important objective of the organization, which also had as one of its objectives the promotion of co-operative research between its member States (though this was not the primary aim of the organization).

Meanwhile the United Kingdom proposed to the countries which signed the Convention of Geneva that they should form a European organization with the aim of building a launcher of heavy satellites, using a British rocket named Blue Streak as the first stage. The Blue Streak rocket was initially intended to serve as a strategic missile, but technically it was judged inadequate for military purposes. The British Government therefore proposed its use for civil scientific purposes, by transforming it into a launcher of heavy satellites by the addition of two stages which were to be built by the European partners.

This project was accepted by France, after some hesitation, and put forward officially by the two countries.

The examination of this proposal took place in the Council of Europe in 1961. A diplomatic conference was held in London at the end of 1961; it led to the drawing up of a convention providing for the creation of an organization for the preparation and construction of space launchers. Only six European countries were interested in the project of the launchers. The organization (ELDO) was formed in April 1962.

A short time later, Europe was faced with a new problem. The possibilities which were opening up in the field of the future exploitation of satellites for telecommunications made it necessary to hold full-scale negotiations at a world level. The United States of America, which had taken the lead in this field, proposed a formula for agreement which in fact ensured this country a preponderant position in a world system.

In May 1963, fifteen European States joined in a European Conference on Telecommunications by Satellite (CETS). This conference settled on an agreed case to be presented by the European countries at the time of the discussions which preceded the Agreement of Washington in 1964 on the building of a world system of telecommunications by satellites. The Washington Agreement had a provisional character and was to be reviewed in 1969. Since 1964, CETS has been endeavouring to lay down a programme of common action to ensure that Western Europe plays a more important role in the world system. It had no budget available to undertake technological development, but national efforts have developed alongside the activities of the international organizations. Several countries have decided to start programmes to build satellites for telecommunications from their own resources or within the framework of bilateral agreements.

This then was the situation in 1968. On the one side three separate European space organizations had been created. Two of them, which had at their disposal international budgets and secretariats, had encountered serious internal difficulties during 1965 and 1966.

These difficulties, concerned with budgets, techniques and organization, had led to serious crises and necessitated the summoning on several occasions of ministerial conferences to discuss the whole range of problems which faced Europe in this sector, and to try to co-ordinate the available resources and define a common long-term space policy.

Thus, at the same time there exist a large number of national space programmes in Europe, usually of small or moderate scope, which overlap technically; yet the future avenues of Europe's space technology are not explored under any international scheme or programme.

The large number of organizations and the absence of any concerted industrial policy have clearly brought Western Europe to an impasse in its space programme.[1]

Example of nuclear research in Europe

Europe's experiences in the nuclear field show many features similar to what has happened in the space programme.

Directly after the Second World War, scientists of various countries, more especially European physicists, consulted each other about the possibilities of reducing the lead gained by the United States in the nuclear field, by providing Europe with equipment on a large enough scale for the study of high-energy physics.

It was within the framework of Unesco that scientists studied the possibility of creating a European centre of fundamental science in the field of high-energy physics. Following an intergovernmental conference, called by Unesco in 1952, a number of European countries signed a convention creating a European Organization for Nuclear Research (CERN). This convention came into force in 1953.

CERN established near Geneva a large laboratory where theoretical studies and fundamental research in nuclear physics are undertaken.

At the same time, governments began to be concerned with the problems —most of them scientific—which must be solved in order that nuclear fission may become a source of energy that can be put to practical use and become economically profitable for industry, naval construction and public utilities. These problems concern the design and building of reactors, their

1. The conference at Bonn in November 1968 made strenuous efforts to straighten matters out.

fuel, the handling of radioactive materials, the separation of fission products, and the disposal of radioactive waste.

A report published under the responsibility of M. Louis Armand had in fact just pointed to the conclusion that there was a danger for Europe of a shortage of the traditional sources of energy and therefore a need to develop nuclear energy to alleviate it.

On the basis of this report, the OEEC[1] decided to entrust to a commission of scientific experts the task of exploring the possibilities of action in the field of nuclear energy open to those European countries which were members of the organization. The proposals of this commission resulted in the establishment in 1957 within OEEC of a permanent organ, the European Nuclear Energy Agency (ENEA), with the mission of 'promoting the development of the production and use of nuclear energy for peaceful purposes by the participating countries, through co-operation between these countries and harmonization of the measures taken on the national level'.

Thus ENEA fosters co-operative research and development work on experimental reactors under a flexible scheme open to the countries which wish to participate in such programmes.

Euratom was established in 1958 in a different context, that of an economic integration between the six European countries working to form the European Common Market. Here, therefore, the initiative came entirely from the governments and the determining factors were of a political nature.

The aims were: (a) to develop nuclear research, especially in the field of reactors, and to publicize the results widely; (b) to achieve a common policy for the supply of fissile materials.

Euratom's programme of research has been carried out partly in four common centres of research, and partly by means of research or associate-ship contracts negotiated between the organization and national institutions such as research centres, universities or industrial enterprises.

Euratom has clearly suffered from disagreements among its members about the advisability of conducting a nuclear armaments policy. This political slant has hampered both the fissile material supply of the research centres and their integration, as well as that of the nuclear industries in the European Economic Community.

However, setbacks have also arisen from the fact that certain problems were under-estimated. In particular, it was believed that in a first phase nuclear research could develop in government centres with many different orientations, and that the industrial phase would be organized within a competitive framework so that all the industries of all the member countries could bid to all the electric power networks of the community.

We know today that programmes for the construction of power stations must be finalized at as early a stage as possible and that it is at this level that a European community is most necessary. If this necessity had been

---

1. Organisation for European Economic Co-operation, which later, in 1962, was to change into the Organisation for Economic Co-operation and Development (OECD) by admitting certain non-European countries: United States of America, Canada and Japan, and later still, Finland and Australia.

realized, no doubt the first thing done would have been to concert government orders leading to the choice of a limited number of branches to be developed. A consortium of industries of several countries (if possible all of them) could then have been set up for each branch. Research in government centres could then have been focused on the requirements of technological development, and could have found its place in a commercially viable policy.

At the end of 1968 the situation in Europe was as follows. For reactors of the first generation, several national firms were commercializing reactors of American design on their own initiative.

For reactors of the second generation, for which R & D was in progress within the framework of ENEA and Euratom, lack of interest on the part of users and constructors for these branches made their future problematical.

Finally, for reactors of the third generation, the 'breeder' and fast reactors which are expected to be exploited commercially in the 1980s, there was a duplication between the French national programme and a joint programme of the Federal Republic of Germany and the Benelux countries. There was an inclination towards industrial rivalry, which itself constituted a threat to co-operation.

Instead of creating a new spirit by the formation of European teams of research, a large part of the Community's budget has been used to subsidize national programmes in member countries. Besides, the problem of uranium-325 had not been solved: in particular, even assuming that this industry did succeed in getting 'off the ground', how was one to ensure the supply of fissile material to the reactors built and exported by European industry?

Many people in Europe have used this situation as an argument for speaking of the failure of European nuclear policy and for criticizing the part played by contemporary institutions, and the forms that co-operation has taken in this sector.

Even the formula of CERN, which has been considered as a success by all its members, and as an example to follow in the field of co-operation, has been in danger of being rejected. Difficulties have arisen when it appeared necessary to increase its capacity for carrying out experiments, and therefore to increase its financial resources. Thus, the first proposal to build a large accelerator of 300 GeV did not receive unanimous support of the partners.

Lesson to be drawn from West European co-operation

Ten years of co-operation in Western Europe make it possible to isolate the essential cause of the difficulties which have been met and the lack of efficiency which springs from the absence of a sufficiently large area of agreement between partner countries on the objectives and long-term prospects of their common action: the initiatives which have been successively taken in this field have not been part of an over-all strategy of economic development which would be of benefit to all the partners.

Hitherto, co-operation has been confined to the stage of research and experimental development, which obviously requires a large outlay of money. In fields like atomic physics and space, which are closely involved with industrial activity, no provision has been made to achieve a common effort for exploitation of the results of the research which was undertaken in collaboration. In other words, up till now international scientific co-operation has been envisaged in the perspective of a 'policy *for* science'. Co-operation has developed without an over-all plan, as new needs gradually become apparent. Pressure (either governmental, or scientific) has led to the establishment of a number of institutions; each of these aims to achieve certain limited objectives, but there has been no effort to obtain a coherent policy among them.

The most obvious result of this evolution is the multiplicity and diversity of the organizations which exist today; their fields of activity overlap, and the list of member countries varies widely from one organization to the other.[1]

It is therefore the absence of a really common policy which is the reason for the precarious situation of international scientific programmes. During these last few years it has become clear that without commitments entered into for a sufficiently long period, and without a broad enough vision of the goal of co-operation, the programmes are always in danger of being held up, especially when unanimous decisions have to be taken and one or other of the countries involved is suffering from financial instability.

And in the absence of a common industrial policy, it has proved impossible to resolve the major problem of the relations between the international programmes and the national programmes. The national industries push on towards the development of the national programmes, which are their own preserve, and this pressure encourages unnecessary duplication.

So it comes about that the expenditure of all the European countries on international programmes in space and nuclear research hardly reaches the total of expenditure of the separate countries on their national programmes in the same field. In these two sectors some large countries even devote to their national programmes much bigger sums of money than they devote to co-operation. Thus co-operation has not replaced national action; nor can one honestly declare that it has represented a complement to national action since in obvious cases it has simply duplicated it.

Random expenditure does not lead to favourable economic results, except in the particular circumstances where a civilization finds itself carried forward on a wave of success. This is clearly not the case of Western Europe in the second half of the twentieth century. Thus, lack of satisfactory economic results gained from scientific expenditure, undertaken by the States in order to overcome the problems of the second phase of industrialization, rouses bitter comments on all sides. The smallest partners, which have no national programme, suspect their larger neighbours—whose economic size is less inadequate and whose enterprises are still in the race—of turning solely to

1. The following is the composition of the European organizations for nuclear and space activities: Euratom contains the six countries of the European Common Market; ELDO, five countries of the Common Market plus the United Kingdom; ESRO, ten countries; CERN, thirteen countries; ENEA, eighteen countries; CETS, eighteen countries.

their own advantage and their own profit the results of research and benefits of an infrastructure which was paid for by all the partners. The reply is that in the present state of efficiency of the European set-up it is a heavy burden on the larger nations to have to pay both the expenses of the European programme (undertaken as a duty of solidarity) and those of the national programme (undertaken in a spirit of industrial efficiency).

From all this it can be concluded that common action in some fields of research can certainly appear to be a promising scheme, but also that research is no longer an isolated activity, which finds in itself its own justification. Thus, in the absence of a genuine common policy in industry, capable of integrating scientific effort within a prospect of economic development favourable to all partners, common activity in research is in danger of coming to an abrupt halt.

Another difficulty springs from the need for a national organization to adapt itself to the existence of international scientific activities.

When there is no consistent national policy, there is a danger that international action will have an emasculating influence on the national scientific potentials. This is why some countries have avoided being associated with co-operative projects, fearing to diminish their national scientific manpower, which is already small, to the benefit of international laboratories. The fact that talented people are tempted to work in international organizations may certainly cause a kind of 'brain drain'. There may be no compensation for the country supplying those specialists, if there is no possibility of the economic and cultural results of the scientific activity in question getting back to their home country.

Governments are equally careful to exercise better supervision over contributions from international resources to their national laboratories. The international funds thus distributed are in reality national resources passing through the channel of international organizations under the form of contracts or subsidies to national centres. This international financing can introduce distortions in the allocation of resources to a national R & D system; it may involve an expensive and unprofitable administrative loop through channels where direct responsibility gets lost.

The need for each country to have a consistent national science policy is nowhere more evident than when international decisions have to be taken. Insufficient co-ordination or even rivalry between ministerial departments within a country are clearly reflected in the inconsistency of the positions which this country is forced to adopt in the international organizations. The multiplicity of these organizations increases the danger of inconsistency, for unless proper co-ordination is achieved at the national level, the positions adopted by the same country in one organization may not be in accord with those adopted in another. It is hardly possible that the sum total of the positions thus taken up by eight or ten countries in ten or twelve organizations can be considered as a consistent European policy of development based on science.

THE FUTURE OF CO-OPERATION

The problems inherent in international scientific co-operation only assumed a definite shape once they became familiar and once the objectives of this co-operation and its implications appeared more clearly.

CERTAIN PRINCIPLES OF INTERNATIONAL SCIENTIFIC CO-OPERATION

International scientific co-operation must be integrated in a 'development policy' accepted by all the partners; it must spring from the same conception of the role of science in the process of economic, social and cultural development.

Therefore the primary justification of co-operation must be to make it possible for the countries concerned to pass through the stages of development more rapidly than they could have done by relying on their own resources.

International science policy cannot be separated from an international economic, social and cultural policy, which must determine both clear and well-defined objectives for common action in the field of research, and also ensure that the results are published and applied in all the participating States.

When international scientific co-operation is viewed in this light it appears that it should be based on four fundamental principles.

The first of these principles is that the co-operation must benefit all the partners. If it does not do so, there is no certainty that it will continue. This interest of the partners ought not to consist entirely in the right of access to the new knowledge produced by co-operative research. Co-operation in research is, in fact, only one means among many whereby each country can encourage the application of science to the benefit of society: countries must also practise such encouragement right the way through to the application of science and technology to economic development. The international interest may indeed be superior to the sum of national interests. It is theoretically possible to imagine a supranational authority having enough vision to concentrate exclusively on carrying out programmes whose only use was for the future of the whole community, or which only gave concrete advantages to those of its members who were furthest advanced in the field in question. However, no international organization could carry out such a policy without disintegrating.

The second principle derives in some measure from the first. It enjoins that each partner in the co-operation obtain his just reward in return for the contribution that he makes. Naturally this reward need not consist of expenditure made within the country concerned or in contracts placed with its industries or laboratories, but in the achievement of the objectives themselves: if the goal is progress in industry, agriculture or public health, it is essential that the international activity should have a direct and profound influence on the real problems in industry, agriculture and public health of each participating country and that these should receive actual dividends in the shape of new jobs, increased production, better health, etc.

The third principle could be expressed in the following manner: the co-operation of member countries must extend to the whole of a policy, and not be confined to individual programmes or projects. The experience of Western Europe has shown that in the absence of a genuine industrial association between the partner nations, countries have been tempted to demand that the allotment of contracts by the various European organizations (Euratom, ESRO, ELDO) should represent a return made in strict proportion to their contribution; these should be given to the industries of the member countries to construct apparatus for research and experiment; in a second phase they practise a system of 'choosing their meal *à la carte',* that is they demand that each member country should have the option of refusing to finance projects which do not fit in with its own programme. Attempts to make this system work have imperilled the very existence of co-operation, since each country has found excellent reasons for opting out of many projects; as a result, financing has become too heavy a burden to be supported by a reduced number of participants, and they have therefore been abandoned. So it seems that the only prospect of stability is in programmes with sufficient scope and continuity to allow for a fair reward for the contributions and for equitable returns to be obtained from a group of long-term projects. It follows from this that the group of the countries in partnership must be sufficiently homogeneous in its needs, and include, if possible, countries which are all at the same phase of their development.

Finally, a fourth principle is that States ought not to undertake in co-operation what they can do equally well alone. The field proper to co-operation is that in which the expenditure is too large for the individual potential of the member countries: i.e., large apparatus, large laboratories, the developments of big technology. For all the subjects of research which do not demand big concentrations of resources, the more flexible forms of co-operation—e.g. exchange, concerted action, without a common international treasury or international bureaucracy—offer fewer dangers of failure.

FACTORS GOVERNING THE EFFECTIVENESS OF CO-OPERATION

Recent experience has shown clearly that effective co-operation implies all the following stages: an objective inquiry into national and international policies and potentials; a common political will for co-operation; a knowledge of the actual needs to which the co-operation must respond; determination of long-term objectives to ensure sufficient continuity in the programmes; an agreement on the forms of co-operation most suitable to attain these objectives. This is why the following factors govern the effectiveness of co-operation.

Information

Co-operation must be founded on an objective knowledge of the national and international positions of the partners. It is necessary that countries should be kept informed on their scientific and technological potentials and the national programmes of their partners, as well as about their channels of political decision, so that they can take all these into account when they determine their line of conduct.

The knowledge of the resources that each country can contribute to the common effort precludes the possibility of founding the co-operation on false premises, or premises insufficiently thought out. It is also essential that information from international organizations should be constantly circulated, to avoid the possibility of new international initiatives duplicating activities already existing at this level.[1]

The needs

Co-operation must be a response to precise needs, felt as such by all partners. Needs are generally indicated by gaps in the activities carried on separately by each country from its own resources. The needs will obviously be more or less clearly defined according to whether they are felt in fundamental research, applied research or experimental development.

The objectives

These must be defined as a function of the available means and the needs which they have to meet.

The objectives must be realistic, taking into account the resources which each partner is prepared to contribute to a common venture. That is, these objectives must be neither too limited nor too ambitious. Clearly the problem for each partner is to assess the advantage of all kinds that it can derive at the national level from its participation in the prosecution of international objectives. The judgement of a member country must be based on a cost-analysis of the possible profits, including a comparison with the alternative choices—which are: either, no project at all, or else, a national project.

Objectives must be sufficiently long-term to ensure the continuity of international action. The field of science is by definition a field which pledges the future.

Some examples may illustrate what may constitute the objectives of international scientific co-operation:

1. Purely scientific and cultural objectives: to assemble sufficient resources to make possible experimental work with a high threshold of efficiency (satellites for experiment and scientific observation, particle accelerators, experimental reactors).
2. Political objectives: to ensure for a group of countries an independent supply of nuclear fuel, or to ensure the independence of a system of space television for cultural or geopolitical ends; the ability to compete in supplying advanced equipment and consumer goods to the world market, or independence in exploiting natural resources such as the sea.
3. Economic objectives: these include objectives capable of having a direct or indirect influence on the economy of all the partners. The following

---

1. In the space field, an inventory of the activities and projects of member States and international organizations was not undertaken until 1966, when ELDO and ESRO were already in a state of crisis.

may be put among the direct objectives: (a) organizational objectives, which aim at adapting the industrial organization of the partners by forming international industrial units capable of competing with any others and by creating a unified internal market of the size which is necessary for this; (b) commercial objectives, which aim at opening up new markets to the industries or agriculture of the partner countries.

4. The objectives of scientific public service: under this heading can be classified services of a scientific character which benefit all the research workers of industry, the universities and the State in the partner countries. A special case of this is the integrated and automated systems of information processing in science and technology.

5. Social objectives: among these are the activities which directly benefit the whole of a population, such as the research into, or government services dealing with, the protection of health, measures of standardization, the exploitation of new sources of food supply, and finally all the measures which aim at developing and improving education. An example would be the organization of post-graduate courses open to a whole group of countries.

FORMS OF CO-OPERATION

Once the objectives have been determined, one has still to determine the most suitable form of co-operation.

The traditional form of the pooling of resources has raised the greatest number of difficulties in these last few years and seems only to be justified by necessity in certain fields of fundamental or applied research, namely, when programmes require large teams from different disciplines working on experimental apparatus or prototypes of large size.

In all other cases, such pooling of resources does not appear to be essential; co-operation has sometimes proved more efficient when it has taken the form of concerting the national initiatives and co-ordinating the national programmes, and sometimes by the simple exchange of information.

So far as the traditional form is concerned, experience has shown the dangers involved in a multiplicity of organizations involved and lack of flexibility in their functioning. Hence it is better to concentrate efforts towards a concerted, integrated programme *in a single institutional entity with a single budget*. The proliferation of financial treaties and financial protocols, which is clearly the result of the diversity of the partners, leads rapidly to a jungle of procedures which jeopardizes the efficiency and continuity of the action undertaken.

These last require a programme of co-operation relying on a sliding plan of reasonable duration (possibly five years), to be readjusted each year—and extended for one year—according to new needs and prospects.

The 'open club' which allows individual countries to withdraw from part of a programme is a more satisfactory solution than to have a large number of institutions, each of which has a different composition and can be obstructed by the veto of one of the partners. But the open club involves the danger of the partners 'choosing their meal *à la carte'*, which destroys the common will. A minimum programme to which all the member States are firmly committed is indispensable to sustain co-operation.

THE POLITICAL WILL TOWARDS CO-OPERATION

It is not enough to determine the objectives and formulate the programmes that are to achieve them. International action should furthermore be supported by a political will towards co-operation shared by all the partners. This is an essential condition of co-operation from which all the rest springs, and without which all co-operative enterprises are doomed to failure.

The political will to achieve an objective in collaboration with one's partners implies a consistent national policy and a clear insight into the limits of national action.

An inconsistent national policy betrays itself by an inefficient organization of the allocation of internal resources, and this condition also has repercussions at the international level.

CONCLUSION

The recent experience of Western Europe shows that international policy cannot make up the deficiences in national policy, and that organization at the international level cannot serve as an excuse for faults in national organization. National weaknesses are magnified at the higher level; they cannot incorporate to form a better alloy. International co-operation is therefore truly profitable only to those nations which have shaped and applied their own policy of development based on science, but which have come up against the obstacle of the inadequacy of their own resources and their own size, and wish to overcome this obstacle by forming an association for specific ends with partners who have the same political outlook in world affairs.

Such a conception of co-operation will perhaps seem unambitious to those readers who hoped for a daring vision in the last chapter of this essay, especially considering that it appears under the patronage of Unesco. Be that as it may, account should be taken of two important restrictions which modify the foregoing conclusions on co-operation.

The first concerns the programmes of scientific public service and the social programmes; these should be given the widest possible scope without posing any conditions. By definition, public service is meant to be extended to all. It constitutes the most effective instrument of international mutual aid.

The second concerns the research programmes of the specialized organizations of the United Nations System. Participation in these programmes ought not to be, and indeed cannot be, confined to countries which are in the same phase of development; neither can there be any question of applying to them the criterion of unity of aim and political outlook, nor that of equality of contributions and benefits. Such programmes must continue to prosper—if only because they are the token of a universal will for progress by co-operation in the splendid scientific adventure of our age.

But it is essential that countries which encounter similar problems—or at least problems which, irrespective of the particular country concerned, pose themselves in comparable terms—should learn to co-operate with each other in smaller groups in order to resolve those problems by means of science.

Conclusions

The scientific and technical revolution at first appeared to mankind as a *phenomenon,* a process of nature in which men found themselves caught up by some inner compulsion or some accident of history.

But as it becomes applied in the life of nations, science gradually loses this character of exteriority. First it was accepted, then taken for granted, and finally passionately desired by the majority of the nations of the earth, both for the promise which it holds out and for the co-operative endeavour which it becomes. For it is a deep-seated characteristic of man's nature to explore his environment and to face its hazards either alone or in a collective venture. This exploration and this involvement have assumed giant proportions; the efficiency of the scientific approach has multiplied the credits as well as the debits, the promises as well as the dangers.

This is the collective venture of our age; it calls on every people, every enterprise and every individual to be involved in one way or another. The implications are: firstly, an effort of selection and preliminary preparation of the objectives (dealt with in Chapter 2 of this work); secondly, a will for efficiency in work and action, which expresses itself by the dissemination into every sphere of human activity of the idea of *technological innovation*—an idea which originally was only current in industry.

Choosing the objectives, and creating the technologies which make it possible to achieve those objectives, are thus the two essential stages of *modern social behaviour.* This will be called *policy* when we speak of a government, *strategy* when we speak of an enterprise or of some group of men united for action or creation, *involvement* when we are speaking of an individual.

In all three cases the scientific attitude imposes its criteria of rationality and efficiency, and it is this which coherently unites man's efforts, despite the tensions and the oppositions or the duplications which arise. So universal is this social behaviour that (save for some small pockets of humanity not yet caught up in the movement—countries which for this very reason we dare to describe as backward) everyone today, in all parts of the globe, can now

216

understand the action which men undertake, and the means employed in such action. The result is that *intersubjectivity* has become very great in the sphere of scientific knowledge and its application. The flows of 'technology transfer' are stronger than ever before. The transfer takes place with great rapidity when it is made between individuals or groups who are specialists in the same field or have had the same experience of science. As between research worker and research worker, doctor and doctor, technician and technician, the transfer is in fact instantaneous, whatever cultural or political barriers may separate their communities, provided their standard of education and activity is comparable.

This characteristic of the scientific age makes it possible for every nation on earth to come into contact with the most advanced findings of science and the most elaborate technologies in the world. It is enough for this purpose that two conditions be fulfilled: firstly, that in each of the special fields there is a sufficient number of scientists and technologists, and that these men should be actively engaged in advanced research in their own field; secondly, that these same men should be linked to the whole of their national community by strong cultural and functional bonds which enable them to have a positive influence on the daily life and guiding decisions of this community. When these conditions are not present, the nation is cut off from the great world movement and so follows its own technological evolution at its own pace, which is bound to be slower than the rest of the world's, unless its economy is taken in hand by outside organizations, whether private or public. In the last case it will only enter the modern world in a subordinate capacity, and so without utilizing the scientific approach on which the decisions of these organizations are founded. This is why the necessity for a direct connexion between the nations and the world of science and technology is seen to be of the first importance by governments. The kind of action which results has been described under the term 'policy *for* science'.

The next stage is more ambitious; it aims to introduce the scientific approach into the process of government decision-making, and by this very act to put science and technology at the service of the objectives of the nation. This stage we have called 'policy of development based *on* science'.

We have attempted to describe how this stage can be reached and passed through, and what demands it makes on government action. The demand that is the most important in its consequences is undoubtedly that of a correct choice of objectives, following upon a clear appreciation of the actual position of the country and the level of economic development which it has reached. No imitation of organizations or attitudes can be a substitute for this thinking. On the contrary, as we have pointed out, imitation can be damaging, since the kind of response which the situation demands differs completely according to whether the country is embarking on its first phase of industrialization, or is deeply involved in it, or is starting or preparing its second phase, or again is coming to grips with the consequences of the latter.

For this paramount reason the present essay has not offered the reader any panacea, and indeed only a few prescriptions. Perhaps all the same it has provided him with a framework for reflection, which can be useful when analysing the position of his own country and the decisions of his own government.

The reader will have noticed in passing that the relative successes and failures of the other nations in their attempts at a policy based on science have more often been historical accidents than deliberate choices. He will also have remarked that the driving power of progress has sometimes shown itself in unexpected ways. This happens in any process of exploration. But the scientific approach prescribes that, from this fact, one should draw rational conclusions which are capable of guiding the action more efficiently in different places and at different times. This was the aim of the present essay, though we are conscious that we have not altogether succeeded.

Finally, we hope that some readers will ask themselves the question which haunts our generation: 'How long will it be before there is a rational approach and therefore a policy of development based on science which can be conceived and applied on a world scale?' Their impatience is justified, and will be still unsatisfied when they shut this book. However, let them consider that if the scientific approach should blossom out, as it must do in the near future, into a world-wide civilization that is genuinely new, each collective unit, each group or enterprise will have its own strategy of progress based on science, since it will not be exclusively at the world-wide level that such strategy will have come into existence. The stage of national policies is therefore not only a necessary stage, but a strong foundation for the building of the future. Indeed, in the majority of countries the national science policy has still to take shape and substance, while in others it is even now engaged in seeking its ways and its means in the midst of obstacles. It is only right that we should not lose sight of these obstacles; but it is no less right that from time to time we should lift up our eyes unto the hills and dream. This at any rate—perhaps only occasionally, perhaps only furtively—is what the authors of the present essay have attempted to do.

218

Bibliography

ARROW, K. J. Economic welfare and the allocation of resources for invention. In: Universities—National Bureau, *The rate and direction of inventive activity: economic and social factors.* 1962.

——; CAPRON, W. M. Dynamic shortages and price rises: the engineer-scientist case. *Quarterly journal of economics,* May 1959.

AUGER, Pierre. Scientific co-operation in Western Europe. *Minerva,* vol. I, no. 4, summer 1964.

BENNETT, William B. *The American patent system.* Baton Rouge, La., Louisiana State University Press, 1943.

BENOIT, E. *Europe at sixes and sevens.* Columbia University Press, 1961.

BERNAL, J. D. *Science in history.* London, 1965.

BRYANT, Samuel W. The patent mess. *Fortune,* September 1962.

CENTRAL ADVISORY COUNCIL FOR SCIENCE AND TECHNOLOGY. *Technological innovation in Britain.* 1968.

COLLETTE, J. M. *La recherche, développement en Grande-Bretagne.* (Cahiers de l'Institut des Sciences Économiques Appliquées, Paris, série T, no. 2.)

CONSEIL NATIONAL DE LA POLITIQUE SCIENTIFIQUE (CNPS).[1] *Recherche et croissance économique, I.* Bruxelles, 1965.

——. *Recherche et croissance économique, II.* Bruxelles, 1968.

COX, Donald W. *America's new policy makers: the scientists rise to power.* Washington, D.C., Chilton Books, 1964.

DAUMAS, Maurice. Esquisse d'une histoire de la vie scientifique. *Histoire de la science.* Paris, Gallimard, 1957. (Encyclopédie de la Pléiade.)

DEDIJER, S. Measuring the growth of science. *Science* (Washington), vol. II, no. 16, 1962.

DÉLÉGATION GÉNÉRALE À LA RECHERCHE SCIENTIFIQUE ET TECHNIQUE. La politique de l'État en faveur du développement industriel des résultats de la recherche. *Le progrès scientifique,* no. 97, juin 1966.

DENISON, E. F. *The sources of economic growth in the United States and the alternatives before us.* New York, Committee for Economic Development, 1962.

——; POULLIET, Jean-Pierre. *Why growth rates differ: postwar experience in nine western countries.* Brookings Institution, 1967.

DEWHURST, J. H. *America's needs and resources: a new survey.* New York, Twentieth Century Fund, 1965.

DJERASSI, Carl. A high priority? Research centres in developing nations. *Bulletin of atomic scientists,* January 1968.

1. Since 1968, the secretariat of the National Science Policy Council has taken the name of the Department for the Programming of Science Policy of the Prime Minister of Belgium.

DOWNIE, J. *The competitive process.* 1958.

DUPRE, Stéfan; LAKOFF, Sanford A. *Science and the nation. Policy and politics.* Washington, 1962.

EDWARDS, A. *Investment in the European Economic Community.* New York, Praeger, 1964.

EWELL, R. H. Role of research in economic growth. *Chemical and engineering news* (Washington), vol. 33, no. 29, 1955.

FAVOREU, L. Un contrat administratif de type nouveau? Les conventions de recherche de la DGRST et de DRME. *Actualité juridique, Droit administratif,* 20 septembre 1965.

FOURASTIÉ, Jean. *Le grand espoir du XXᵉ siècle.* Paris, Presses Universitaires de France, 1949.

——. *The causes of wealth.* Illinois, 1960.

FREEMAN, C.; YOUNG, A. *The research and development effort in Western Europe, North America and the Soviet Union.* Paris, OECD, 1965.

FUCHS, Victor R. *Productivity trends in the goods and service sectors, 1929-1961: a preliminary survey.* New York, National Bureau of Economic Research, 1967.

GABRIEL. *The international transfer of corporate skills.* Cambridge, Mass., Harvard University Press, 1967.

GALBRAITH, John K. *The new industrial state.* Boston, Mass., Houghton Mifflin Co., 1967.

GERRITSEN, J. C. The problems and methods of financing scientific and technical research. *Meeting of Experts on the Role of Science and Technology in Economic Development, Unesco, Paris, 11-18 December 1968.* (SPS.18.)

GRANICK, D. *The European executive.* New York, 1962.

GUSTAFSON, E. Research and development, new products, and productivity change. *American economic review,* May 1962.

HANSTEIN, M. H. D.; NEUMANN, K. *Die ökonomische Analyse des Technischen Niveaus der Industrieproduktion.* Berlin, Die Wirtschaft, 1965.

HAYASHI, Y. The problems and methods of financing scientific and technical research in Japan. *Meeting of Experts on the Role of Science and Technology in Economic Development, Unesco, Paris, 11-18 December 1968.* (SPS.18.)

HEMPTINNE, Y. de; PIGANIOL, P.; VU CONG, L. National development, technological innovation and research programming. In: *The promotion of scientific activity in tropical Africa. Transactions of the Symposium on Science Policy and Research Administration in Africa, Yaoundé (Cameroon), 10-21 July 1967.* (SPS.11.)

HERTZ, D. B. La recherche-développement considérée comme facteur de production. *Économie appliquée* (Paris), 1961.

HIRSCH, S. The U.S. electronics industry in international trade. *National Institute economic review,* no. 34, November 1965.

HOFFMANN, F. *Forschung in der Amerikanischen Industrie.* Köln, 1922.

HUMPHREY, Hubert H. The need for a department of science. *Annals of the American Academy of Political and Social Science,* January 1960.

HUVELIN, P. Les marchés d'étude et de recherche passés par les administrations publiques, régime français. In: *Aspects juridiques de la recherche scientifique.* Faculté de Droit de l'Université de Liège et La Haye, Nijhoff, 1965.

HUYGENS, Christian. *Correspondance, Œuvres complètes,* vol. IV. La Haye, Nijhoff.

INSTITUT POUR L'ÉTUDE DES MÉTHODES DE DIRECTION DE L'ENTREPRISE (IMEDE). *The Swedish investment reserve system. A special feature in economic policy.* Lausanne, 1964.

JANTSCH, E. *Technological forecasting in perspective.* Paris, OECD, 1967.

JEFIMOV, A. N. Die Rolle der Verflechtungsbilanz bei der Optimierung Volkswirtschaftlicher Proportionen. *Wirtschafts-Wissenschaft* (Berlin), no. 10. 1965.

JEWKES, J.; SAWERS, D.; STILLERMAN, R. *The sources of invention.* London, 1958.

JOHNSON, E. A.; STRINER, H. E. The quantitative effect of research on national economic growth. *International Conference on Operation Research, Aix-en-Provence (France), 1961.*

KEEZER, Dexter M. The outlook for expenditures on research and development during the next decade. *American economic review, papers and proceedings,* May 1960.

KINDLEBERGER, Charles P. *Economic development.* New York, McGraw-Hill, 1965.

KING, A. International scientific co-coperation—its possibilities and its limits. *Impact,* vol. IV, no. 4, 1953.

KING, James E. *Science and rationalism in the government of Louis XIV.* Baltimore, Md., Johns Hopkins Press, 1949.

KUSICKA, H.; LEUPOLD, W. *Industrieforschung und Ökonomie.* Berlin, 1966.

LAYTON, C. *Transatlantic investment.* Paris, Atlantic Council, 1968.

LECERF, D. Major research and development programmes as instruments of economic strategy. *Impact,* vol. XVII, no. 2, 1967.

L'ESTOILE, H. de. Choice of criteria for a research and development strategy. *Meeting of Experts on the Role of Science and Technology in Economic Development, Unesco, Paris, 11-18 December 1968.* (SPS.18.)

LYONS, Sir Henry George. *The Royal Society, 1660-1960.* Cambridge University Press, 1944.

MANSFIELD, E. Rates of return from industrial research and development. *American economic review,* May 1965.

MARLEWICZ, M. Problems and methods of financing scientific and technical research (in the socialist countries belonging to the Council for Mutual Economic Aid). *Meeting of Experts on the Role of Science and Technology in Economic Development, Unesco, Paris, 11-18 December 1968.* (SPS.18.)

——. Principles of the system of financing scientific and technical research. *Meeting of Experts on the Role of Science and Technology in Economic Development, Unesco, Paris, 11-18 December 1968.* (SPS.18.)

MATTHEWS, R. C. O. Contribution of science and technology to economic development. *Meeting of Experts on the Role of Science and Technology in Economic Development, Unesco, Paris, 11-18 December 1968.* (SPS.18.)

MINASIAN, J. R. The economics of research and development. In: Universities—National Bureau, *The rate and direction of inventive activity: economic and social factors.* 1962.

NELSON, R. R. Uncertainty, learning and the economics of parallel research and development efforts. *Review of economics and statistics,* November 1961.

——; PECK, M. J.; KALACHEK, E. D. *Technology, economic growth and public policy.* Washington, 1967.

——; PHELPS, E. S. Investment in humans, technological diffusion, and economic growth. *American economic review,* May 1966.

NEUMEYER, F. Employees' rights in their inventions: a comparison of national laws. *International labour review,* January 1961.

ODHIAMBO, T. R. East Africa: science for development. *Science,* 17 November 1967.

ORGANISATION FOR ECONOMIC CO-OPERATION AND DEVELOPMENT (OECD). *The overall level and structure of R & D efforts in OECD member countries.* Paris, 1967.

——. *Industrial research associations in France, Belgium and Germany.* Paris, 1965.

——. *Government and technical innovation.* Ministerial Meeting on Science, Paris, 1966.

——. *The measurement of scientific and technical activities: proposed standard practice for surveys of research and experimental development.* (Now incorporated in: *The Frascati manual,* rev., Paris, 1970-71.)

——. *International scientific organisations.* Paris, 1965.

——. *Reviews of national science policy: Belgium.* Paris, 1966.

——. *Reviews of national science policy: United States.* Paris, 1968.

——. *Report on the organisation of scientific research in France.* Paris, 1964.

——. *Economic growth 1960-1970.* Paris, 1966.

——. *The impact of science and technology on social and economic development.* Paris, 1968.

——. *Gaps in technology between member countries.* General report of third Ministerial Meeting on Science, Paris, 1968.

——. *Resources of technical and scientific personnel in the OECD area.* Paris, 1963.

OECD. *Science, economic growth, and government policy.* Paris, 1963.

——. *A study of resources devoted to research-development in OECD member countries in 1963-1964.* Paris, 1967.

PAIGE, R. D.; BOMBACH, G. *A comparison of national output and productivity of the United Kingdom and The United States.* Paris, OECD, 1959.

PECK, M. J. Science and technology. In: R. E. Caves *et al., Britain's economic prospects. 1968.*

PERROUX, François. *Le progrès économique.* Paris, Presses Universitaires de France, 1965.

PEVTCHIN, G. Les marchés d'étude et de recherche passés par les administrations publiques, régime des États-Unis. In: *Aspects juridiques de la recherche scientifique.* Faculté de Droit de l'Université de Liège et La Haye, Nijhoff, 1965.

PHELPS, E. S. Models of technical progress and the golden rule of research. *Review of economic studies,* 1966.

PIERRE, A. La recherche scientifique en URSS. *Industrie* (Bruxelles), no. 6, juin 1961.

PIGANIOL; VILLECOURT. *Pour une politique scientifique.* Paris, 1963.

PRICE, Don K. *Government and science: their dynamic relation in American democracy.* Oxford University Press (Galaxy Books), New York, 1954.

QUINN, J. B. National planning of science and technology in France. *Science,* 19 November, 1965.

——. Technological competition, Europe vs. U.S. *Harvard business review,* July-August 1966.

——. Scientific and technical strategy at the national and major enterprise level. *Meeting of Experts on the Role of Science and Technology in Economic Development, Unesco, Paris, 11-18 December 1968.* (SPS.18.)

ROBERTS, Edward B. The dynamics of research and development. Address to the National Security Industrial Association, 3 November 1965.

ROSE, H. The rejection of the WHO Research Center. *Minerva,* vol. IV, no. 3, spring 1967.

RUIC, Neil P. The case for going to the moon. Part VII: The case for technological transfer. *Industrial research,* March 1965.

SAINT-HIPPOLYTE, M. de. L'écart de productivité entre les États-Unis et les pays industrialisés d'Europe. *Le progrès scientifique,* no. 120, juin 1968.

SALOMON, J. J. International scientific policy. *Minerva* (London), summer 1964.

SALTER, W. E. G. *Productivity and technical change.* Cambridge University Press, 1960.

SAMUELSON, R. J. Israel: science-based industry figures large in economic plans. *Science,* 24 May 1968.

SCHMOOKLER, Jacob. *Invention and economic growth.* Cambridge, Mass., 1966.

SCHUMPETER, A. *The theory of economic development.* Cambridge, Mass., Harvard University Press, 1949.

SERVAN-SCHREIBER, J.-J. *The American challenge.* London, Hamish Hamilton, 1968.

SHELL, K. Towards a theory of inventive activity and capital accumulation. *American economic review,* May 1966.

SHONFIELD, A. *Modern capitalism.* Oxford University Press, 1965.

SILBERSTON, A. The patent system. *Lloyds Bank review,* April 1967.

SOLOW, R. M. Investment and technical progress. *Mathematical methods in the social sciences.* Stanford, 1960.

SORBIERE. Discours à l'Académie de physiciens, 3 avril 1963. *Académie des sciences, 1917,* tome 1964.

STEINITZ, K. Zeitfaktor und Effektivität der sozialistischen erweiterten Reproduktion. *Wirtschafts-Wissenschaft* (Berlin), November 1965.

STONEHILL, A. *Foreign investment in Norwegian enterprises.* Oslo, Statistisk Sentralbyra, 1965.

STOVER, Carl F. *The government contract system as a problem in public policy: the industry-government aerospace relationship.* Stanford Research Institute, January 1963.

SZAKASITS, G. D. Various approaches to the problem of the integration of scientific and economic plans into general planning. *Meeting of Experts on the Role of Science and Technology in Economic Development, Unesco, Paris, 11-18 December 1968.* (SPS.18.)

——. La planification prospective des capacités de recherche-développement. *Közgazdasági Szemle* (Budapest), no. 9, 1962.

——. *Scientific research and economic development.* Budapest, Akedémiai Kiadó, 1965.

TOMS, M.; HÁJEK, M. Determinants of economic growth and integral productivity. *Politika ekonomie* (Prague), no. 10, 1966.

UNESCO. *Directory of international scientific institutions.* 2nd ed. Paris, 1953.

——. *Current trends in scientific research.* Paris, 1961.

——. International Conference on the Organization of Research and Training in Africa in Relation to the Study, Conservation and Utilization of Natural Resources, Lagos (Nigeria), 28 July to 6 August 1964. *Scientific research in Africa. National policies, research institutions.* Paris, 1966.

——. Conference on the Application of Science and Technology to the Development of Latin America, Santiago, Chile, 13-22 September 1965. *Final report.*

——. *Guidelines for the elaboration of national science policy studies.* Paris, 1965. (Unesco/NS/ROU/85.)

——. *La politique scientifique et l'organisation de la recherche scientifique en Belgique,* Paris, 1965. (SPS.1.)

——. *Science policy and organization of scientific research in the Czechoslovak Socialist Republic.* Paris, 1965. (SPS.2.)

——. *National science policies in countries of South and South-East Asia.* Paris, 1965. (SPS.3.)

——. *Science policy and organization of research in Norway.* Paris, 1966 (SPS.4.)

——. *World directory of national science policy-making bodies.* Vol. 1: *Europe and North America,* 1966; vol. 2: *Asia and Oceania,* 1968; vol. 3: *Latin America,* 1968.

——. *Principles and problems of national science policies.* Paris, 1967. (SPS.5.)

——. *Structural and operational schemes of national science policy.* Paris, 1967. (SPS.6.)

——. *Science policy and organization of research in the U.S.S.R.* Paris, 1967. (SPS.7.)

——. *Science policy and organization of research in Japan.* Paris, 1967. (SPS.8.)

——. *Science policy and the organization of scientific research in the Socialist Federal Republic of Yugoslavia.* Paris, 1968. (SPS.9.)

——. *National science policies of the U.S.A. Origins, development and present status.* Paris, 1968. (SPS.10.)

——. *Contribution of Unesco to stage 1 of the world plan of action for the application of science and technology to development.* Paris, 1968. (Unesco/NS/ROU/155.)

——. *The problem of emigration of scientists and technologists.* Paris, 1968. (Unesco/NS/ROU/158.)

——. *Proceedings of the symposium on science policy and biomedical research organized by CIOMS with the assistance of Unesco and WHO, Unesco House, Paris, 26-29 February 1968.* Paris, 1969. (SPS.16.)

——. *Symposium on brain research and human behaviour, Paris, 11-15 March 1968. Final report.*

——. *Meeting of science policy experts preparatory to the Conference of Ministers for Science of the European Member States of Unesco, Bucharest (Romania), 23-30 April 1968. Final report.*

——. *Science policy and its relation to national development planning. The application of science and technology to the development of Asia: basic data and considerations.* 1968.

——. *Guidelines for the drafting of a 'National Summary' analysing the present situation of, and future prospects for, science policy in a European Member State of Unesco.* (Unesco/NS/ROU/163.)

UNESCO. *Activities of Unesco in the field of promoting scientific and technical co-operation.* Paris, 1968. (Unesco/NS/ROU/172.)

——. *Unesco's activities in the field of national science and technology policies.* Paris, 1968. (Unesco/NS/ROU/169.)

——. *Second Meeting of Directors of the Councils for Science Policy and Research of the Latin American Member States, Caracas (Venezuela), 10-17 December 1968. Preliminary report.*

——. *The role of science and technology in economic development.* Paris, 1970. (SPS.18.)

——. *Use and conservation of the biosphere. Proceedings of the Intergovernmental Conference of Experts on the Scientific Basis for Rational Use and Conservation of the Resources of the Biosphere, Paris, 4-13 September 1968.* Paris, 1970. (Natural resources, X.)

——. Conference on the Application of Science and Technology to the Development of Asia, New Delhi (India), 9-20 August 1968. *Science and technology in Asian development.* Paris, 1970.

——. *Bilateral institutional links in science and technology.* Paris, 1969. (SPS.13.)

——. *Manual for surveying national scientific and technological potential.* Paris, 1970. (SPS.15.)

——. *The promotion of scientific activity in tropical Africa. Transactions of the Symposium on Science Policy and Research Administration in Africa, Yaoundé (Cameroon), 10-21 July 1967.* (SPS.11.)

UNION OF INTERNATIONAL ASSOCIATIONS. *Yearbook of international organizations.* 12th ed. Bruxelles, 1969.

UNITED NATIONS. *Science and technology for development. Report on the United Nations Conference on the Application of Science and Technology for the Benefit of the Less Developed Areas, Geneva, 4-21 February 1963 (UNCSAT).* 8 vols. Transfer and Adaptation of Technology for Developing Countries. New York, 1963. (United Nations doc. E/C D/31.)

UNITED STATES CONGRESS. JOINT ECONOMIC COMMITTEE. *Higher unemployment rates, 1957-1960, structural transformation or inadequate demand.* Washington, D.C., Government Printing Office, 1961.

USHER, D. The welfare economics of invention. *Economica,* August 1964.

VAN HOORN, J. *Régime fiscal de la recherche et du développement technique.* Paris, OECD, 1961.

VILLECOURT. *Formes de coopération dans les problèmes de politique scientifique, Séminaire de Jouy-en-Josas, 19-25 février 1967.* Paris, OECD, 1967.

WEINBERG, A. Criteria for scientific choice. *Minerva,* winter 1963.

WELLES, John G. et al. *The commercial application of missile/space technology.* Denver Research Institute, University of Denver, 1963.

WELLES, John; WATERMAN, Robert. Space technology. Pay-off from spin-off. *Harvard business review,* July-August 1964.

WILLIAMS, B. R. Research and economic growth—what should we expect? *Minerva,* autumn 1964.

WOODWARD, F. N. *Structure of industrial research associations.* Paris, OECD, 1965.